BIOTECHNOLOGY IN AGRICULTURE SERIES

General Editor: Gabrielle J. Persley, International Service for National Agricultural Research (The Hague, Netherlands), and Project Manager, World Bank/ISNAR/ACIAR/AIDAB Biotechnology Study.

For a number of years, biotechnology has held out the prospect for major advances in agricultural production, but only recently have the results of this new revolution started to reach application in the field. The potential for further rapid developments is however immense.

The aim of this new book series is to review advances and current knowledge in key areas of biotechnology as applied to crop and animal production, forestry and food science. Some titles will focus on individual crop species or groups of species, others on specific goals such as plant protection or animal health, with yet others addressing particular methodologies such as tissue culture, transformation or immunoassay. In some cases, relevant molecular and cell biology and genetics will also be covered. Issues of relevance to both industrialised and developing countries will be addressed, and social, economic and legal implications will also be considered. Most titles will be written for research workers in the biological sciences and agriculture, but some will also be useful as textbooks for senior-level students in these disciplines.

Editorial Advisory Board:
P.J. Brumby, formerly of the World Bank, Washington DC, USA.
E.P. Cunningham, Trinity College, University of Dublin, Ireland.
P. Day, Rutgers University, New Jersey, USA.
J.H. Dodds, International Potato Center (CIP), Peru.
J.J. Doyle, International Laboratory for Research on Animal Diseases, Nairobi, Kenya.
S.L. Krugman, United States Department of Agriculture, Forest Service.
W.J. Peacock, CSIRO, Division of Plant Industry, Australia.

Titles Available:
1: Beyond Mendel's Garden: Biotechnology in the Service of World Agriculture *G.J. Persley*
2: Agricultural Biotechnology: Opportunities for International Development *Edited by G.J. Persley*
3: The Molecular and Cellular Biology of the Potato *Edited by M.E. Vayda and W.D. Park*

Titles in Preparation:

Biotechnology of Perennial Fruit Crops *Edited by F. Hammerschlag
and R. Litz*

Rice Biotechnology *Edited by G.S. Khush and G. Toenniessen*

Plant Genetic Manipulation for Crop Protection *Edited by A. Gatehouse,
V. Hilder and D. Boulter*

Barley: Genetics, Molecular Biology and Biotechnology
Edited by P.R. Shewry

Beyond Mendel's Garden

Biotechnology in the Service of World Agriculture

by

Gabrielle J. Persley

C·A·B International

for

The World Bank

Australian Centre for International Agricultural Research
Australian International Development Assistance Bureau
International Service for National Agricultural Research

For
Sir John Crawford
a great Australian and internationalist,
who showed me the light on the hill.

———————————————

C·A·B International
Wallingford
Oxon OX10 8DE
UK

Tel: Wallingford (0491) 32111
Telex: 847964 (COMAGG G)
Telecom Gold/Dialcom: 84: CAU001
Fax: (0491) 33508

British Library Cataloguing in Publication Data
Persley, Gabrielle J. (Gabrielle Josephine)
  Beyond Mendel's garden : biotechnology in the service of world agriculture.
  1. Agriculture. Applications of biotechology
  I. Title
  630

  ISBN 0-85198-682-X
  ISSN 0960-202X

Typeset in Britain by Leaper & Gard Ltd
Printed and bound in the UK by Bookcraft (Bath) Ltd

Contents

Foreword

Modern biotechnologists, or genetic engineers, are becoming the new partners of agricultural scientists. They see new ways of tackling old problems, through the application of biotechnologies ranging from the long-established commercial use of microbes and other living organisms to the genetic engineering of microorganisms, plants and animals.

The World Bank supports agricultural research through loans and credits to national governments and grants to the international agricultural research centres, particularly those sponsored by the Consultative Group on International Agricultural Research (CGIAR). There is a continuing need for new technologies to accelerate rural development, especially for resource-poor farmers and smaller countries.

In 1988, the World Bank commissioned a study to assess the potential of biotechnology to contribute to increased agricultural productivity, and to identify the socioeconomic, policy and management issues that may affect its successful application. The study is co-sponsored by the World Bank, the International Service for National Agricultural Research (ISNAR), the Australian Centre for International Agricultural Research (ACIAR), and the Australian International Development Assistance Bureau (AIDAB). The study is co-financed by the World Bank and the Australian Government. Additional support for country studies is being provided by the Government of The Netherlands and the United Nations Development Programme (UNDP).

The aims of the study are to identify the opportunities and constraints in the use of biotechnology to solve agricultural problems in developing countries, and to develop strategies for increased investments in biotechnology by national agricultural research systems and international development agencies. Biotechnology is composed of a continuum of technologies ranging from long-established and widely used technologies through to emerging technologies for the genetic engineering of plants and animals. The study is considering possible points of intervention along the continuum that would contribute to development in different countries. The study culminated in a symposium on: Agricultural Biotechnology: Opportunities for International Development, held in Canberra, Australia, in May 1989, in association with the mid-term meeting of the CGIAR. Participants came from several Asian, African and Latin American countries, the international agricultural research centres and several bilateral and multilateral development agencies.

The CGIAR has acknowledged the importance of biotechnology as a component of the research and development process. It is a subject being monitored by the CGIAR Technical Advisory Committee, the Boards of Trustees and the Directors-General of the individual IARCs. In 1989, a CGIAR Task Force on Biotechnology (BIOTASK) was established to raise awareness amongst all components of the CGIAR system of the issues involved in biotechnology, and to provide information and advice on selected subjects. The publication of this volume (and its companion volume *Agricultural Biotechnology: Opportunities for International Development*) is therefore a timely contribution to the consideration of these important issues.

Wilfield P. Thalwitz
Chair
Consultative Group on
International Agricultural
Research

Preface

In May 1988 I was given a special assignment by the Australian Centre for International Agricultural Research to act as Project Manager for an international, and inter-agency study on agricultural biotechnology. I was based at the International Service for National Agricultural Research (ISNAR) in The Hague, and worked closely with Peter Brumby and Tony Pritchard in the Agriculture and Rural Development Department of the World Bank in Washington.

CAB International took an early interest in the project, since that group was planning to launch a new book series on 'Biotechnology in Agriculture' and was interested to act as publisher for the monographs from this study. Throughout 1988–89 a number of reports and documents were issued by ISNAR and ACIAR, and a technical report summarising the main findings was prepared for publication by the World Bank. One of our early goals, however, was to synthesise the material in a way that would provide an overview of the future role of biotechnology in Third World agriculture for policymakers and research managers who were not specialists in molecular biology. The result of this early planning with the publisher, and the hard work of contributors to the study in many countries, is this book.

This volume discusses the potential benefits of, and constraints to, the application of biotechnology to agriculture in developing countries, and the various policy issues confronting national agricultural research systems and the international development community. Areas covered include the likely socioeconomic impact of biotechnology on world trade and economic development; opportunities for the application of biotechnology in crops, forestry and livestock production; the issues involved in the management of intellectual property; the need for suitable regulatory processes to govern the safe release of genetically engineered organisms into the environment; the opportunities for private/public sector co-operation; the future role of the international agricultural research centres in facilitating access to the new biotechnology in their client countries; and the policy, educational and management issues affecting national agricultural research systems as they seek to integrate biotechnology into their agricultural sectors. Summaries of several commodity analyses are included, which consider the possible applications of new technologies to solve problems in some major tropical crops. These (and other) crops may be termed 'orphan commodities', ones on which there is little investment in modern biotechnology in industrialised countries, but which are important food and/or cash crops in developing countries. Included also is a summary of ten country case studies, and a discussion of some future options for international development agencies in supporting agricultural biotechnology.

A companion monograph is being published by CAB International which contains the text of the 31 commissioned papers for the study. It is entitled *Agricultural Biotechnology: Opportunities for International Development*.

This volume does not require the reader to have a specialist knowledge of biotechnology. It is intended to provide information to those who have responsibilities for the investment of human and financial resources in biotechnology relevant to the Third World, such as managers in national agricultural research systems, and policymakers concerned with rural development.

Acknowledgements

This publication is a synthesis of information provided in 31 commissioned papers, 10 country studies, and several invited papers presented at a symposium on 'Agricultural Biotechnology: Opportunities for International Development' held in Canberra, Australia, in May 1989. The valuable contributions of all authors are gratefully acknowledged. Their names are listed in the references section, and the commissioned papers are being published by CAB International in a companion volume. The participants at the Canberra Symposium provided a spirited discussion of issues, and I have attempted to reflect some of their views in this volume.

Peter Brumby and Tony Pritchard were the World Bank staff members overseeing the study, and their many valuable contributions to the text and wise advice are gratefully acknowledged. Clive James and Val Giddings provided additional material and their contributions are much appreciated. Jock Anderson, Robert Auger, John Barton, Rob Bertram, Simon Best, Mark Cantley, Peter Dart, Jack Doyle, Greg Gibbons, Kerri-Anne Jones, Allen Kerr and Nancy Millis provided suggestions for improvement of parts of the text which were most helpful.

Skilful secretarial assistance was provided by Arlene Slijk-Holden, ISNAR, Pam van den Heuvel, ACIAR, and Jocelyn Barra, The World Bank. Reg MacIntyre undertook the technical editing of the manuscript and his contributions are warmly acknowledged. The continued encouragement of Don Mentz, Denis Greenland and Tim Hardwick at CABI is always welcome.

Several members of the sponsoring organisations provided encouragement throughout the course of the study: Dr W. David Hopper, Dr Michel Petit and Dr S. Bargouti, The World Bank; Dr G.H.L. Rothschild, ACIAR; Dr Bob Dun, Dr Richard Manning, Mr Denis Fitzgerald and Mr Hilary Coles, AIDAB; Dr Christian Bonte-Friedheim, Dr Howard Elliott and Dr H.K. Jain, ISNAR; Mr Alexander von der Osten, Executive Secretary of the CGIAR; and Prof. J.R. McWilliam, Canberra. Their support and guidance is warmly acknowledged.

I am also grateful to Professor Allen Kerr, Waite Institute, University of Adelaide, and Dr Jim Peacock of the CSIRO Division of Plant Industry, Canberra, for stimulating my interest in genetics, and for their continuing support and encouragement.

Biotechnology policy options: a prologue*

W. David Hopper
Senior Vice-President, World Bank, Washington DC

We need to view biotechnology both as a source of immense
promise for the future, and as a source of potential difficulty ...

We are in a period of extraordinarily rapid change in modern biology. This
is notwithstanding the concerns expressed by some that while we have
generated much enthusiasm and many claims for biotechnology, can we
really live up to them? The claims exceed our grasp at the present moment,
but not by much.

It is not too many years ago that I was discussing molecular biology
during a visit to Israel. The point made by the scientists at that time was
that eventually we would take the double helix, lift it up and put some new
genes into it, and put it back together again. There was scepticism
expressed by many at that conference as to whether it would be that easy.
In general, there was a feeling that maybe it was, but it would be the next
century before we manage to make the necessary breakthroughs. The
progress we have made since the 1960s and 1970s leads me to believe that,
in fact, things are moving faster than anticipated initially, and that we have
made many significant steps forward.

As I look at the policy options for international development agencies
supporting biotechnology for agriculture, I begin with the first option,
which is not to do it at all. This approach is one that made a strong impres-
sion on me at one point in my career. I was asked by a president of a
university what would happen if he said no to the request for the first
computer at the university. My comment was: 'You can say no this year,
you have to say it again next year, and the year following. If you say yes,
you will never have another opportunity to say no. It's in here and from
then on the computer drives that decision in the future.'

Do we have any choice when we look at the policy options? Can we say

---

*Keynote Address at a Symposium on 'Agricultural Biotechnology: Opportunities
for International Development', Canberra, Australia, 25–27 May 1989.

one of the policy options is that we will not support it? I think this is an untenable option. *I think that it is inevitable that those of us who are concerned with international agricultural research are going to be supporting biotechnology in the future.*

The issues are clear. Malthus has been reigning now amongst economists and ecologists for about 150 years and he has been proven wrong, but we still operate with the Malthusian paradigm as to what our future is going to be. We have 5.2 billion people, according to recent population estimates, which is expected to be 10 billion in about 30 years. That is about a 2.4% growth in population per year.

The Green Revolution has underpinned part of the population increase in the developing countries. It has permitted worldwide growth of food production to keep pace with population growth. The Green Revolution really was only a method of raising tropical food production to about the same level as had been attained in the temperate zones through the application of research. When we began the process of creating the International Rice Research Institute in the late 1950s, temperate food production levels were two to three times that of the tropical level. The question was, could tropical food production levels increase to approximately the same level as the temperate level? They could, and did, thanks to the efforts of the brilliant scientists who worked on the major crops, particularly rice and wheat.

We now find that we are running at about the same level of productivity in the tropics, at least in the research plots, as we are operating in the temperate zone. Gradually the developing countries are putting in the infrastructure that is necessary to support a much higher productivity in their agriculture. Africa lags a bit behind, but those that are concerned with agriculture in Africa know that the potential is there and that we are slowly uncovering the techniques for handling it. Unless there is a very significant advance in productivity (a significant advance far greater than I see evident from the present data from the CGIAR centres, and others engaged in tropical agricultural research), I do not see us being able to beat the Malthusian proposition by the year 2000 or 2030. The next significant advances must come from genetic engineering.

The developing countries, where the significant population growth is, must be our partners in this effort to increase food production. It is not going to be a case of the industrialised countries taking responsibility for providing adequate food to the developing countries. It is going to be a case of how do we build a partnership between the industrialised and the developing countries that will transfer to the developing countries new technologies and, more importantly, the means of acquiring those new technologies that are relevant to solving their future food production problems?

Since I believe that most of these new technologies are going to be based on genetic engineering, the developing countries must become our partners in biotechnology. I return therefore, to the proposition that the first policy option is that we must do it – and the real question is how do we build the partnerships to effect the new relationship, a research relationship, a transport of information relationship between the developing countries and the industrial nations? Fostering that partnership logically rests with the international development agencies.

The biotechnology revolution is a major change in how we approach agriculture and how we approach agricultural science. The change is incremental to the scientists who are engaged in it, but for those who are on the outside, it is a colossal set of advances, and it is a total change in how we are approaching science, and therefore how we approach (in my view at least) the policies that we need for assisting the development of science in the Third World.

Biotechnology draws from virtually all branches of science. It seems to spawn, almost weekly, some new discipline or cross-disciplinary field. That integration I find fascinating, in the sense that we can no longer talk about just agriculture or biotechnology, we have to talk about a broad range of biology and its interrelation with other branches of science, such as physics and chemistry. Biotechnology involves a multi-disciplinary approach to science. Donors have to recognise that as they approach biotechnology.

Biotechnology also blurs the nice distinctions with which we were comfortable, as to whether we were financing basic or applied science, and whether we were doing pure research or problem-oriented research. For development agencies we knew that we supported applied, but not basic, science, and that we were only interested in problem-oriented research and not basic research. It seems to me now that we have much more of a *seamless web*, linking the basic scientists with the problems in the field much more closely than before. The distinction that we as donors like to have, that we are not financing fundamental science, that we are really focused on problems and the problem orientation of research, seems to disappear in this effort.

This seamless web links farmers to the laboratories in ways that they have not been linked before. It raises questions that have not been raised before; it requires that development agencies be much more sensitive in today's world as to how they structure the partnerships in biotechnology between the industrialised and the developing nations, and the research institutions in each.

We must also be sensitive to the skills that are required and the staffing development that is needed. We must be sensitive to the fact that the biotechnology revolution opens up some opportunities to do things anywhere in the world, that do not have to be site-specific. For example, we can take a rice genome and work on it at Cornell University and make that research relevant to the lowland tropics. Conversely, the specificity with which we can examine genes, opens up real opportunities to tailor-make new plants, and new livestock for specific environments. It is a double game, of both broader adaptability of certain research, and of greater specificity in our targets.

While traditional plant breeding awaited the playing of the roulette wheel, and some were better at playing roulette than others, much genetic manipulation can now be done in the laboratory. In order to make the laboratory work applicable to the field, the productive environment where we wish to use the newly bred plant needs to be much better understood. The enhanced power that we can give to the plant breeders to tailor-make varieties for specific environments raises again the question of the partnership, and the sensitivity in this relationship between the biotechnology laboratories and our partners in the developing countries.

Among the policy options, exactly what are we going to do with technical assistance? What do we mean by this in relation to biotechnology? Is it just going to be a question of taking someone who has an advanced degree in molecular biology and moving him/her to a developing country for a few years? Alternatively, are we going to try to structure a different kind of technical assistance? Let me suggest it is going to have to be a different kind of technical assistance.

The relationships that we need to establish in biotechnology are between equal partners, in industrialised and developing countries. They are different relationships to that which we are used to in conventional 'technical assistance' programs. They are ones we are going to have to work at and discover how these relationships can be forged. They are going to be much more collegiate relationships. They are going to be between persons, and groups that are well trained; maybe the individuals will be different in age and experience but that is really the only difference. The partnerships must be true two-way relationships for them to work. As donors, we are going to have to recognise the two-way nature of the relationship, and change many of our rules accordingly. It will be necessary to have two-way trips by the collaborating scientists; it will be necessary to provide training opportunities; it will be necessary to continue the capital build-up in developing countries; and it will be necessary to commit our funds to long-term support for our partner institutions in the developing countries, not for two years, three years, or even five years, but for ten years and beyond. We are also going to have to get our political masters to understand why these modifications to our rules are necessary to foster the desired collegiality amongst scientists.

The international development agencies are also going to have to settle their policy alternatives. We are going to have to take a look at just how much of the globe we want to tackle; how much of our work is global, and how much is going to be specific; and how much, as donors, we are really prepared to examine the shifts in comparative advantage that could be brought about by biotechnology. What do we do when we change comparative advantage between countries, and between commodities? How much are we prepared to struggle with the potential of biotechnology for that change?

Science today seems more dominated by patent questions than it is dominated by the old traditions of science. There is presently an intense and sometimes bitter discussion in progress in the General Agreements on Tariffs and Trade (GATT) and the World Intellectual Property Organization (WIPO) as to what exactly are intellectual property rights, and who controls them. It is not scientists who determine this, it is lawyers. They say that economists have become lawyers in this age, and mathematicians have become economists. It may be that scientists will have to become lawyers as well.

Let me look briefly at some other items that need to feed into a policy framework for biotechnology. First, I believe we need to bring much more specificity to the overall goals of biotechnological development as we seek to work with the developing countries. Precisely what are we going to do in biotechnology? In visiting the international agricultural research centres, and some individual countries it seems to me that biotechnology for the sake of biotechnology seems to hold a very central position. I would rather ask the questions, 'What do we want biotechnology to do?' 'Can we define how we are going to do it?' There are those grand five Ws of the newspaper profession that need to be gone through: *Who? What? Why? When? Where?* There is also *How?* and *How much?*

I would like to see us begin to develop with our developing country partners, and amongst our development agencies, a strategy for biotechnology, a set of plans which says, 'This is where it is going to go; These are the countries.' For example, Brazil, India and China have advanced capabilities in this field. Some others are probably substantially behind in biotechnology at present. As we approach this subject, I think that it might be useful to be as coldly objective about the strategy as we want to be about the biotechnology itself. This is implicit in any banker's statement, and that is: "What are the costs relative to the benefits?' 'What do we see and hope for here? And what do we see as our course?'

In biotechnology I believe this is not easy to do, the reason being that the field is fairly new. I can give you some good estimates of costs/benefits in traditional plant breeding, as to how long it takes for those genetic roulette wheels to line up. Indeed we had such estimates when we began discussing the creation of IRRI for the Rockefeller Foundation back in the late 1950s. In biotechnology, the breakthroughs are still coming rather fast, and relatively unpredictably, and it is not easy to line up our hopes with some rough estimates on costs. As a Bank representative, I would like to see it a little more firmly placed as to: (1) what we think we can do, (2) where we think we can go and (3) what the costs are likely to be. I also would like to see much more discussion with the developing countries as to what the goals are going to be.

For example, in Africa, many countries are concentrating their research efforts on food production for domestic purposes. Yet many countries are dependent for foreign exchange on a small number of export commodities, and most are in the midst of a commodity crisis. There may well be a significant contribution from biotechnology to increase the export competitiveness of selected commodities and even to foster diversification into new export commodities. Chile is an outstanding example of a country that identified the potential of new export commodities and developed the technology to enable it to increase its export competitiveness. In the case of Chile, it was the export of fruits and vegetables to the North American market. There may be other similar opportunities.

Another important consideration in a policy discussion on biotechnology is that of training. What kind of training do we provide for our partners? We seek to provide them with the best, and I think that is what we have to insist on here. We need to address the questions of what type of training is suitable? How many people require training? What incentives are required for them to return to their home country? The training questions are going to have to be addressed far more specifically than we have done in the past.

In the same way, the questions of institution-building are going to have to become paramount to what we do. The key question on institution-building is how to provide the long-term institutional support and its linkages to the research groups in the industrialised and newly industrialised countries and in the more advanced of the developing countries, that I described earlier.

Critical to the institution question, is the related issue of information and documentation. The provision of timely documentation on advances in biotechnology is an essential part of effective partnerships.

Another important institutional issue is the provision and maintenance of suitable equipment. The maintenance of equipment is one of the major problems in developing countries that we have not succeeded in solving even today with more conventional equipment.

As donors we are also going to have to look at the policy we are going to follow on conditionality, not just the question of competitiveness, but on the issue of patents, and on the issue of legal restrictions, particularly environmental regulations. Are these to be part of our conditionality? We are somewhat draconian in what we do in the USA, and I sense in some of our regulations, what I would call an anti-science bias. It is a major bias. The conditionality question is something that the World Bank struggles with continuously, but it is one that we do not have a significant or agreed policy among donors as we look at biotechnology. One of the major policy questions that we need to look at is environmental regulations, to govern the release of novel organisms into the environment.

The final question that the development agencies are going to have to struggle with, is what kind of research structure and organisation are we going to have? We have heard of the opportunities and constraints for private/public sector collaboration in biotechnology. As representatives of industrialised countries, which of our biotechnology institutions are the appropriate partners for our Third World colleagues? Is it our public institutions? Is it the private offshoot of the public institutions that have been created to deal with patent issues and the like? Is it our private companies? There are codes of conduct for the transnational corporations that are now available through the United Nations. Are these going to be the codes that guide us in biotechnology? These issues need to be resolved, and suitable ways devised to deal with the different types of institutions.

The rationale for agricultural research being in the public sector was because agriculture was a much more competitive industry, and therefore it was not possible for any one farm to gain much advantage from an investment in research because that research would spread very quickly throughout the industry. There would be little profit to capitalise or to capture from that research activity. As we move forward in biotechnology there will be opportunities for profits to be captured. What then is the role that we have as development agencies? Obviously, we are back to the questions of patents and other legal regulations, and their associated policies. I do not have answers for this, but they become an important element of whatever policies we wish to adopt.

With regard to research structures in developing countries, I am not too rigid about this. I do not believe that there is a single ideal research structure. Research structures can become a negative factor in itself. I am not sure that there is an ideal research organisation. I do think that this is where the integration point becomes critical. I think we are going to have to look not just at the support being given to individual research institutes but the support being given to the whole research system. We should also consider not just a national system but the regional system, especially in a region with many small countries. When we talk about regional or national systems in large countries, we are going to have to begin to think about how a country or a region can make most effective use of the capital items such as the equipment and laboratories that biotechnology requires. This leads me to the belief that there is merit in considering a campus approach: a single major centre for advanced research on all commodities (crops, trees, animals, fish and microorganisms). We cannot as development agencies support the equipment base and the training base in each of the institutions that each developing country has for the different commodity groups. This is beyond our resources, and it is not the most efficient use of the available human and financial resources.

There must therefore be a campus approach, a single centre for certain common attributes of science; this would be a centre to which trained people returned and a centre which is supported as a long-term endeavour from both national and international sources. From there it would be essential to work with the individual institutes, which deal with specific commodity groups, and test the products of the biotechnological work. The optimisation of cost becomes a major question for donors and for recipients as we build the partnership between the industrialised countries and the developing countries in biotechnology.

Finally, there are the questions about: (1) the specific targets for biotechnology research; (2) who owns what we have developed; (3) how do we get it disseminated; and (4) if it is produced for a profit return, how do we ensure that a legitimate return is made for those who made the investment?

The policy list is far from complete. These are some of the salient issues as we move forward into examining what are the policy questions for our donor community, in their assistance and support for biotechnology in agricultural research. I believe firmly that we have no choice, that we will go forward. I believe that we can either do it efficiently or we can do it much in the way that we have done things in the past, which is case by case, with each donor acting independently, each following a separate set of policies. In this case I think that the effectiveness, particularly the cost effectiveness of what we undertake, will be substantially lessened from the effectiveness that could be if we sat down and planned this from the beginning. *We need to view biotechnology both as a source of immense promise for the future, and as a source of potential difficulty, as we try to adapt it to the world that still thinks in the way that we approached traditional agricultural research in the past, the way the Ford and Rockefeller foundations started the International Rice Research Institute: 'That was a good model; let's not get too far away from it.' It is not a good model for today, and we* must *get away from it.*

Chapter one:
Introduction

The current excitement about genetic engineering ... stems from the ability of scientists to manipulate and control genes in new ways ...

Introduction

'Biotechnology' has been defined by the Office of Technology Assessment of the United States Congress as 'any technique that uses living organisms, or substances from those organisms, to make or modify a product, to improve plants or animals, or to develop microorganisms for specific uses' (OTA, 1989).

'Biotechnology' includes 'traditional biotechnology', covering well-established technologies used in commercially useful operations. These include the technologies presently used in brewing, biological control of pests, conventional animal vaccine production and many other biotechnological applications.

'Modern biotechnology' encompasses the use of more recently available technologies, particularly those based on the use of recombinant DNA technology, monoclonal antibodies (MCA) and new cell and tissue culture techniques, including novel bioprocessing techniques.

Biotechnology is thus comprised of a continuum of technologies, ranging from the long-established, and widely used technologies, which are based on the commercial use of microbes and other living organisms, through to the more strategic research on genetic engineering of plants and animals.

Jones (K.A. 1990) illustrated this continuum in the form of a gradient of biotechnologies, ranging from the relatively simple technologies for the selection of beneficial strains of microorganisms to the more complex methods of genetic engineering of plants and animals (Fig. 1.1).

This study considered possible points of intervention along the continuum of biotechnology that would contribute to agricultural development in the Third World.

Fig. 1.1   K.A. Jones' (1990) gradient of biotechnologies.

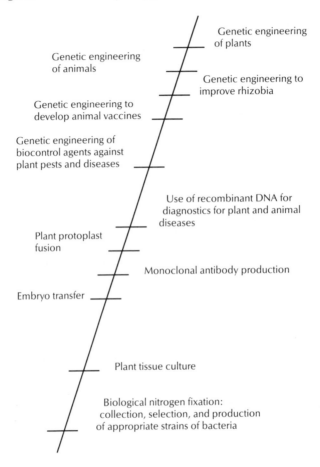

Objectives

The objectives of the study were:

1. To assess the potential of biotechnology (particularly, but not exclusively, modern biotechnology) to increase agricultural productivity, and to contribute to the sustainability of agricultural systems in the Third World.
2. To identify options for investments by international development agencies in the generation and/or application of biotechnology.

Scope of the study

The study considered the following areas:

1. Modern biotechnology achievements in the agricultural sector, the problems likely to be encountered in incorporating these advances into existing practices, and the implications of increasing use of new technologies.
2. Identification of the options available to individual countries to use modern biotechnology within national research systems.
3. Possibilities for collaboration amongst national research groups in developing countries and biotechnology groups in the private and public sector in industrialised countries, the international agricultural research centres and elsewhere.
4. Issues to be addressed by the World Bank in its efforts to advance the use of biotechnology in agricultural research and rural development projects.

Many of these matters are also being addressed by others, and the study drew on available material in assessing the current situation and identifying future opportunities. Thirty one papers were commissioned on selected technical and policy issues, to provide substantive analysis of the scientific possibilities afforded by modern biotechnology for some of the major tropical commodities, and the socioeconomic and management issues that will affect their successful application (Persley, 1989b).

Several country case studies were initiated in countries of different sizes, in different geographic areas and at different stages of economic growth. Their aim was to identify opportunities and constraints for the application of biotechnology in specific countries, and to see if there are any lessons of wider applicability from the comparative analysis. Initial country studies were conducted in the ASEAN member countries (Brunei Darussalem, Indonesia, Malaysia, the Philippines, Singapore and Thailand), India, China, Mexico and Brazil.

New technologies

Historical perspective

The current excitement about genetic engineering and its potential impact on agriculture stems from the ability of scientists to manipulate and control genes in new ways. These recent developments should be seen in historical perspective, as part of the evolution of the science of genetics and biology. A chronology of the development of the science of genetics, and its evolution into modern biology, is shown in Table 1.1.

In 1866, Gregor Mendel, an Austrian monk, observed the inheritance of morphological features in pea plants. Based on these observations, he devised a set of laws to explain the inheritance of biological characteristics. Mendel's basic proposition was that each heritable property is determined by a physical factor, contained within the cells of a living organism. Mendel's laws of inheritance were rediscovered by the scientific community early this century, after a number of other scientists had come to the same conclusion independently. This discovery marked the birth of the science of genetics.

There have been three significant periods in the development of the science of genetics:

1866–1920	*Classical genetics*: Mendel's basic principles accepted, and the existence of genes demonstrated; location of genes on chromosomes demonstrated; linear arrangements and location of genes shown by genetic mapping.
1944–66	*Central dogma*: genes shown to consist of a chemical, deoxyribonucleic acid (DNA); structure of DNA determined; genetic code deciphered.
1971 to date	*Genetic engineering*: development of recombinant DNA techniques which allow genes to be manipulated, and transferred from one species to another, often unrelated, species; genetic engineering of bacteria, plants and animals initiated.

Principles of genetic engineering

Modern biotechnology is based on new techniques in: (1) recombinant DNA technology; (2) monoclonal antibody production; and (3) cell and tissue culture. It is the combination of these three processes that forms the basis of genetic engineering of microbes, plants and animals.

Recombinant DNA technology is a series of enabling techniques for genetic engineering, which allows the manipulation of DNA, the essential genetic material in the cells. Monoclonal antibodies are specific diagnostic tools which allow the rapid detection of individual proteins produced by the cells. Recent advances in cell and tissue culture allow the rapid propagation of genetically engineered cells.

Genetic engineering has both direct and indirect applications in agriculture. The direct applications are concerned with the addition of one gene (or at most a few genes) to a genotype. It is analogous to the situation in conventional plant breeding where a single, dominant gene is added to a particular genotype. The indirect applications are concerned with the availability of more accurate and rapid diagnostic tests for plant and animal

Table 1.1. The evolution of the science of genetics, leading to modern biotechnology.

1866 Mendel postulates a set of rules to explain the inheritance of biological characteristics in living organisms.

1900 Mendelian law rediscovered after independent experimental evidence confirms Mendel's basic principles.

1903 Sutton postulates that genes are located on chromosomes.

1910 Morgan's experiments prove genes are located on chromosomes.

1911 Johannsen devises the term 'gene,' and distinguishes genotypes (determined by genetic composition) and phenotypes (influenced by environment).

1922 Morgan and colleagues develop gene mapping techniques and prepare gene map of fruit fly chromosomes, ultimately containing over 2000 genes.

1944 Avery, MacLeod and McCarty demonstrated that genes are composed of deoxyribonucleic acid (DNA) rather than protein.

1952 Hershey and Chase confirm role of DNA as the basic genetic material.

1953 Watson and Crick discover the double-helix structure of DNA.

1960 Genetic code deciphered.

1971 Cohen and Boyer develop initial techniques for rDNA technology, to allow transfer of genetic material from one organism to another.

1973 First gene (for insulin production) cloned, using rDNA technology.

1974 First expression in bacteria of a gene cloned from a different species.

1976 First new biotechnology firm established to exploit rDNA technology (Genentech in USA).

1980 USA Supreme Court rules that microorganisms can be patented under existing law (Diamond *v.* Chakrabarty).

Cohen/Boyer patent issued on the technique for the construction of rDNA.

1982 First rDNA animal vaccine approved for sale in Europe (colibacillosis).

First rDNA pharmaceutical (insulin) approved for sale in USA and UK.
First successful transfer of a gene from one animal species to another (a transgenic mouse carrying the gene for rat growth hormone).
First transgenic plant produced, using an agrobacterium transformation system.

1983 First successful transfer of a plant gene from one species to another.

1985 US Patent Office extends patent protection to genetically engineered plants.

1986 Transgenic pigs produced carrying the gene for human growth hormone.

1987 First field trials in USA of transgenic plants (tomatoes with a gene for insect resistance).

First field trials in USA of genetically engineered microorganisms.

1988 US Patent Office extends patent protection to genetically engineered animals.

First genetically modified microorganism approved for commercial sale as a biocontrol agent of a plant disease (crown gall of fruit trees in Australia).

1989 Human genome mapping project initiated.

1990 Plant genome mapping projects (for cereals and *Arabidopsis*) initiated.

Source: Adapted from OTA (1984) and Wyke (1988).

diseases, and the addition of new techniques for plant and animal breeding to increase the efficiency of breeding programs.

The types of novel products which are being produced by the application of the new technologies include: enzymes; food additives; biocontrol agents; biofertilisers; animal growth hormones; animal vaccines; diagnostic reagents for plant and animal diseases; new plant varieties; and new animal breeds.

Genetic engineering evolved from an understanding of how cells function naturally, particularly how the genetic material (DNA) codes for the production of proteins essential for the life of the cell. Based on this understanding, other scientists then devised a series of new techniques, collectively called *recombinant DNA technology*, to allow the manipulation of these processes in the cell.

The basic principle of genetic engineering is that genetic material (DNA) can be transferred from a cell of one species to another, unrelated species and made to express itself in the recipient cell. The newly formed DNA in the recipient cell contains both its own, naturally occurring genes, plus the new gene. The new *recombinant DNA* is inserted into the chromosome of the *transgenic plant or animal.* The recipient cell (and organism) are said to be *transformed* by the arrival of the new genetic information (see Fig. 1.2). The key components of genetic engineering are:

1. *Identification and isolation* of suitable genes to transfer;
2. *Delivery systems* to introduce the desired gene into the recipient cells;
3. *Expression* of the new genetic information in the recipient cells.

Rapid progress is being made in refining the techniques that allow a gene from one species to be transferred and expressed in another species. The major limitation is to identify genes which, when transferred with appropriate molecular controls, will confer agriculturally useful traits on the recipient microorganism, plant or animal.

The scientific principles upon which the new technologies are based are described in more detail in Persley and Peacock (1990).

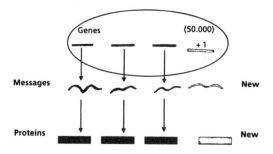

Fig. 1.2   Genetic engineering = gene addition

Chapter two:
Constraints to agricultural productivity

It is important to combine the skills of agricultural scientists ... with the skills of modern biologists who see new ways to approach old problems ...

Introduction

The first step in assessing the usefulness of new technologies to agriculture is to identify the problems which have proved to be intractable to conventional approaches and which may benefit from the application of new technologies. The key questions to be considered in relation to agricultural biotechnology are:

1. What are the major problems to be solved in a country or region, in relation to agricultural productivity?
2. What new products or processes are needed to solve these problems?
3. Do these products and processes exist, or are they being developed elsewhere?
4. If the required products or processes exist elsewhere, are they available for transfer?
5. If the required products and processes do not exist, do they need to be developed specifically in the country or region of need?
6. If the required products and processes need to be developed, what is the most efficient way to have them developed, and where?

The focus in biotechnology thus needs to be on the problems to be solved, and the products and processes needed to solve them, rather than on the technology itself.

In the analysis of what needs to be done, it is important to combine the skills of agricultural scientists and economists who understand the commodity and its needs, with the skills of modern biologists who see new ways to approach old problems. Cell and molecular biologists are the new partners of agricultural scientists not their replacements.

For such new partnerships to work, there needs to be more bridging courses and retraining opportunities to enable agricultural scientists and research managers to understand better the basic principles of modern biotechnology and how to use new techniques (such as new diagnostic techniques, genetic engineering and genetic mapping) in their existing research programs. It would also be valuable for modern biologists to understand more of the agronomy and ecology of the target species and its pests and pathogens. Another resource is the cadre of past graduates in the biological and agricultural sciences. It is important that opportunities be provided for them to upgrade their skills and learn new techniques to complement their research experience. Modern biology should be perceived to be the province of the curious, whatever their age, not merely of the young. *Investors in biotechnology need to assess in individual countries the requirements for bridging courses and retraining programs in biotechnology to facilitate the successful introduction and integration of new technologies into existing R and D programs.*

In plant agriculture, the major route for the application of new technologies will be by the development of seeds of new plant varieties with novel characteristics (such as herbicide tolerance, resistance to specific pests and diseases). The widespread use of these novel characteristics will be dependent on their further incorporation through conventional plant breeding into varieties with suitable agronomic characteristics and stable yields.

Plant breeding is being held back in many countries by institutional and/ or infrastructural barriers. For example, quarantine regulations often preclude timely importations of needed new germplasm. Standardised, simple equipment may be unavailable due to import restrictions. Difficulty in travel may restrict the number of sites in which trials can be placed and thus limit the predictions that can be made about the range of adaptation of new varieties. Government regulations may delay introduction of promising new varieties to farmers. Poor communications can limit use of computer networking to facilitate data collection and analysis. Local-product laws may prevent use of standardised computer equipment and necessitate special programming or even preclude the use of computers. Fluctuating power supplies may hinder the maintenance of medium- and long-term cold storage for breeding seeds (Duvick, 1990).

A further hindrance to the development of conventional plant breeding in many countries is the series of problems faced by private industry in supporting local private-sector plant breeding programs. These problems include outright prohibition of in-country research in some countries, difficulties of importation of proprietary germplasm; proprietary protection often restricts commercial activity to breeding and sale of hybrid crops, which gives a patent-like protection to the owners of the parent lines (Duvick, 1990).

Greatly strengthened plant breeding programs are a necessary prerequisite for the successful adoption of modern biotechnology in developing countries. The International Agricultural Research Centres (IARCs) could play a pivotal role in influencing the development of national capability in plant breeding. Much more needs to be done by the IARCs and international development agencies to strengthen national plant breeding programs.

Commodity studies

Overview

Several commodities important in the Third World have been examined to assess their current constraints to productivity, and the likely availability of new technologies which might be applied for their solution. The crops considered here as examples are banana/plantain, cassava, cocoa, coffee, coconut, oil palm, potato, rapeseed, rice and wheat. The availability of new technologies for each commodity are listed in Table 2.1. The current constraints for each commodity, and the potential biotechnology solutions are given in Table 2.2.

Substantial progress may be expected in the short term (0–5 years) for potato, rapeseed and rice; in the medium term (5–10 years) for banana/ plantain, cassava and coffee; and only in the long-term (10+ years) for cocoa, coconut, oil palm and wheat.

Table 2.1. Availability of new technologies for selected crops.

Crop[a]	New diagnostics[b]	Rapid propagation systems[c]	Transformation systems[d]	Regeneration systems[e]	Time-frame[f]
Banana/ plantain	+	+	—	+	medium
Cassava	+	+	+	—	medium
Cocoa	+	—	—	—	long
Coconut	+	—	—	—	long
Coffee	+	+	—	+	medium
Oil palm	+	+	—	—	long
Potato	+	+	+	+	short
Rapeseed	+	+	+	+	short
Rice	+	+	+	+	short
Wheat	+	+	—	—	long

Notes: [a]Crop: Illustrative examples of crops important as food and/or export crops in the Third World.

[b]Availability of new diagnostics for pests or diseases based on the use of monoclonal antibodies or nucleic acid probes.

[c]Availability of rapid propagation systems to allow the multiplication of new varieties.

[d]Availability of transformation systems to enable new genetic information to be inserted into single plant cells.

[e]Availability of regeneration systems to enable single cells to be regenerated into whole plants.

[f]Likely time-frame for commercial applications of new technology: short-term, now– <5 years; medium, 5–10 years; long, >10 years.

Table 2.2.   Commodity constraints and potential biotechnological solutions.

Commodity	Constraints	Potential biotechnological solutions	Time-frame[a]	Reference
Banana/ plantain	Black Sigatoka bunchy top virus *Fusarium* wilt	new diagnostics host plant resistance	long short long	Dale (1990)
Cassava	high cyanide cassava mosaic virus	new varieties coat protein resistance	long medium	Bertram (1990)
Coffee	coffee rust quality characteristics	host plant resistance new varieties	medium long	Söndahl (1990)
Cocoa	lack of rapid vegetative propagation	novel tissue culture techniques	long	Söndahl (1990)
Coconut	lack of vegetative propagation virus/viroid diseases lethal yellowing disease	novel tissue culture techniques new diagnostics genetic mapping	medium short long	Jones, L.H. (1990)
Oil palm	clonal propagation flowering abnormalities drought susceptibility pests and diseases oil quality	novel tissue culture methods genetic mapping genetic engineering genetic engineering	medium long long long	Jones, L.H. (1990)
Potato	high temperature susceptibility lack of disease-free planting material	genetic mapping new tissue culture techniques and new diagnostics	short short	Dodds and Tejeda (1990)
Rapeseed	oil-quality	genetic mapping	medium	Scowcroft (1990)
Rice	virus diseases	genetic engineering	medium	Toennison and Herdt (1989)
Wheat	fungal diseases virus diseases (barley yellow dwarf)	genetic engineering genetic engineering	long, medium	Larkin (1990)

Note: [a]Short, now or <5 years; medium, 5–10 years; long, >10 years.

The early applications of modern biotechnology are likely to be on crops which are primarily of interest to the industrial world, since this is where the substantial R and D investments are being made. *There is a need for additional investments in biotechnology on commodities important in the Third World, in order to ensure that these commodities also benefit from the application of new technologies.*

Banana/plantain

Banana and plantain (*Musa* spp.) are monocots, and represent the largest fruit crop in the world. They are the least developed of all the major food crops, and would benefit from nonconventional breeding approaches since conventional breeding has so far been unsuccessful in developing new varieties for commercial use. All production is still exclusively from natural selections, despite many years of effort to breed improved banana hybrids suitable for commercial cultivation.

The major constraints to banana/plantain production are diseases, particularly the fungal leaf disease Black Sigatoka (Black Leaf Streak), which is spreading rapidly in Africa and Latin America and causing major yield losses, in both banana and plantain; *Fusarium* wilt, a soil-borne fungal disease; and bunchy top virus, present in Asia and parts of Africa, but not in Latin America.

Plant pathology particularly had benefited from the development of biotechnology and recent advances could be applied to banana/plantain. New pathogen detection techniques based on monoclonal antibodies and DNA probes are being prepared for the improved diagnosis of fungal and viral pathogens of banana/plantain (Dale, 1990).

New genetic techniques are available for investigating variability in organisms. These techniques could be used in the investigation of *Musa* germplasm diversity and pathogen diversity, including the identification of different races of pathogens, such as the *Fusarium* wilt organism. This information could be used in the selection of disease-resistant lines.

There is no known source of resistance to bunchy top virus. It may be possible to genetically engineer banana/plantain by introducing a gene for bunchy top virus resistance based on the viral coat protein gene. The introduction of genes for resistance to the major fungal diseases (Black Sigatoka and *Fusarium* wilt) is a long-term goal as genes for fungal resistance are not presently available for any crop.

Genetic engineering of banana/plantain requires a transformation and regeneration system that is not presently available. Transformation via microprojectiles offers most promise (Dale, 1990). Recent progress on the regeneration of banana from cell protoplasts suggests that routine regeneration systems will soon be established for the crop (Novak *et al.*, 1989).

Cassava

Cassava (*Manihot esculenta*) is a dicot species, whose roots are widely used as a source of starch for human and animal consumption in Africa, Latin America and parts of Asia. It has high yield potential, drought tolerance and an ability to grow on poor soils.

In September 1988, scientists from Latin America, Africa and elsewhere met at the Centro Internacional de Agricultura Tropical (CIAT) in Cali, Colombia, to identify the major constraints to cassava production, which, if overcome by the use of new biotechnologies, would have major impact on the productivity of the crop.

The targets identified by the group reflected a consensus amongst scientists from Latin America and Africa. The African situation had been assessed earlier at a meeting organised by the International Institute of Tropical Agriculture (IITA), Ibadan, Nigeria, and these priorities are taken into account in the following analysis (Bertram, 1990).

The major constraints to cassava production are its: (1) high cyanide content; (2) vegetative propagation system; (3) rapid postharvest deterioration in root quality; and (4) susceptibility to virus diseases, especially African cassava mosaic.

The research goals established the need for:

1. Transformation and regeneration systems (these are required to enable genetic engineering of cassava for desirable traits to occur).
2. Development of true-seed propagation systems.
3. Elimination of cyanogenic compounds.
4. Resistance to cassava mosaic virus through viral-coat protein mediated protection.

Coffee

Coffee (*Coffea* spp.) is a valuable export commodity for several developing countries, with a world market of US$12 billion. The long-term strategy for coffee-producing countries is to increase farmer revenues, stabilise production and deliver consistent quality. To lower farming costs, one has to focus on fertiliser efficiency, disease and pest resistance, and greater efficiency in production. Yield increases will dilute direct and indirect costs but can lower international prices due to overproduction (Söndahl, 1990).

There is a great deal of knowledge accumulated for supporting coffee improvement programs. There are germplasm collections and extensive breeding experience, and new techniques in cell and molecular genetics. Arabica coffee is a self-pollinating species and Robusta coffee is an outcrossing species. These distinct characteristics will determine different breeding methods and will influence propagation schemes.

The following research targets are recommended for coffee improvement (Söndahl, 1990):

1. Increased nutrient uptake and utilisation.

2. Stable resistance to major diseases (such as coffee leaf rust, coffee berry disease).

3. Tolerance to major pests (e.g. leaf miner, nematodes, coffee borer).

4. Tolerance to adverse environmental conditions (e.g. drought, heavy metals).

5. Flavour enhancement.

6. Caffeine levels (high and low).

Cocoa

Cocoa (*Theobroma cacoa*) generates approximately US$2.6 billion in the commodity market, and so it is also an important source of hard currency for many developing countries, including several African countries.

Cocoa is a highly heterozygous species and crosses easily with other *Theobroma* species. The main production limitations in cocoa are related to fungal and viral diseases. There is a great deal of natural variability in commercial plantations and in natural populations in the centres of variability. Selection of disease-resistant individuals and rapid vegetative propagation would have a great impact on cocoa production. Heterosis has been demonstrated in cocoa and so programs to produce true hybrids would increase yield and open up other opportunities such as improved butter quality and flavour enhancement (Söndahl, 1990).

Traditional breeding in cocoa has not been as extensive as coffee. Also, tissue culture techniques are not as advanced in cocoa as in coffee. Presently, no adult cocoa plant has been recovered from an *in vitro* system. Somatic embryogenesis has been reported from immature sexual embryos but no survivors under greenhouse or nursery conditions are available. Similarly, no anther culture, liquid and protoplast cultures have been described for this species. This means that the application of new technologies to cocoa will be a long-term prospect.

The following priority research targets for cocoa are recommended (Söndahl, 1990):

1. Rapid vegetative propagation systems:
  – somatic embryogenesis from nonsexual tissues,

2. Anther culture.

3. Liquid culture and protoplast techniques.

Coconut

The coconut palm (*Cocos nucifera*), is a pan-tropical crop, primarily of the coasts and islands. It is widely grown by smallholders for use as a food and for many other domestic purposes, and as a cash crop. The primary export commodity from coconut is coconut oil, a high-value lauric oil with many industrial uses. The primary exporter is the Philippines.

The coconut industry is facing many problems. These include: (1) the declining productivity of ageing plantations; (2) the lack of adoption of potentially high-yielding hybrids; (3) the lack of a rapid propagation system for elite palms; and (4) several pest and disease problems, some being lethal diseases of unknown aetiology.

Coconut is one of the tropical crops for which there is a potential threat from biotechnology. The development of a temperate oilseed crop which produced lauric oils would be damaging to the export markets for coconut and palm kernel oil, which both command a price premium as lauric oils which have special industrial uses. In order for coconut to retain even its present share of the market, it is important that more effort be put into R and D to improve the productivity of the crop (Persley, 1990a, b).

There is an opportunity for genetic improvement of coconut, since there is much genetic variability from which to select. The major constraint is that there is little understanding of the genetic basis of disease resistance and other desirable traits, such as drought tolerance and oil quality (Jones, L.H., 1990).

The first application of modern biotechnology to coconut is the recently established technique for the culture of coconut embryos, to facilitate germplasm collection and exchange. These techniques have now been established by the International Board for Plant Genetic Resources (IBPGR) and the Institut de Recherche pour les Huiles et Oléagineux (IRHO) in such a way as to be able to be used in the field for the collection of coconut germplasm (IBPGR, 1988). New diagnostic tests for coconut virus and viroid diseases based on the use of DNA probes have also been developed successfully for use in the Philippines and in the Pacific Islands (Randles and Hanold, 1988).

The next required application of tissue culture is the micropropagation of clonal palms from selected elite individuals. This would enable major improvements in yield and quality to be achieved. There have been several reports of the clonal propagation of isolated coconut palms. However, there are no reliable reports of the availability of a rapid, routine method for the clonal propagation of coconut palms (Jones, L.H., 1990).

The application of more sophisticated genetic manipulations to coconut is dependent on the development of reliable, efficient plant regeneration and transformation systems for cell cultures.

Much of the current work on tissue culture is being conducted by private companies and is shrouded in trade secrecy. A concentrated effort should be made to overcome the technical difficulties in coconut tissue culture, and to establish reliable transformation and regeneration systems. This could involve both public and private sector institutions in the precompetitive research to establish the cell and tissue culture techniques to enable the crop to be rapidly propagated, and to be genetically engineered. The competitive stage is to know which elite palms to propagate, and to identify and isolate useful genes to insert into the crop (Jones, L.H., 1990).

There is a need to develop a genetic map of coconut, in order to identify the genetic basis of control of a number of important traits of interest to individual breeding programs. This information would contribute to the identification of desirable parents in directed breeding programs.

The specific technologies required for coconut are:

1. A genetic map of the coconut genome to identify useful traits for plant breeders, such as resistance to lethal yellowing disease and cadang cadang disease.
2. Improved tissue culture systems for embryo culture, and establishment systems for the resulting plantlets.
3. Transformation and regeneration systems for individual cells.
4. New diagnostics for coconut diseases.

Oil palm

Oil palm (*Elaeis guineensis*) is a valuable source of vegetable oil. It is primarily an export crop, originating in West Africa, but now widely grown in Asia and Latin America. The principal exporter is Malaysia.

There is opportunity for the genetic improvement of oil palm, as it is a highly variable crop. Clonal propagation systems have been established to enable the rapid propagation of elite individuals. These systems have been tested on a semi-commercial basis, but have encountered difficulties due to flowering abnormalities in the clonally propagated material (Jones, L.H., 1990).

There is a lack of understanding in oil palm of the genetic basis of control of the characters that would be of interest to modify, such as yield, oil quality, drought resistance and disease resistance. The preparation of a genetic map for oil palm which shows the locations of the genetic controls for the desirable traits would be valuable. This information could be used by plant breeders to direct their breeding programs and greatly reduce the number of palms that have to be carried in large and expensive field trials (Jones, L.H., 1990).

The research targets in oil palm are: (1) drought tolerance; (2) insect and disease resistance; (3) oil quality.

The technologies that need to be developed to meet these targets are:

1. Improved clonal propagation systems, which produce normal flowering plants.
2. Regeneration and transformation systems.
3. A genetic map of oil palm to locate some key traits of interest to plant breeders.

Potato

Potato (*Solanum tuberosum*) is an important crop in both industrialised and developing countries. It originated in the mountains of South America, was taken to Europe and developed commercially there, and is now becoming increasingly important throughout the Third World, both as a cash and a food crop. In monetary terms, it is the fourth most valuable food crop in the Third World, after rice, wheat, and maize.

The present constraints to potato production are:

1. High temperatures which do not favour potato production in lowland areas.
2. The lack of high-quality, disease-free seed tubers.
3. High production costs and risks of crop failure due to climatic hazards, pests and diseases.
4. Postharvest problems in the tropics.

Potato is a dicot species, for which rapid, efficient tissue culture propagation systems and new regeneration and transformation systems have been developed. There is considerable work being done on the genetic engineering of potato in laboratories in the USA and in Europe, since it is one of approximately 20 economic crops that can be genetically engineered.

The International Potato Centre (CIP) in Peru has been able to build on the research being conducted on potato in temperate climates, by the establishment of a potato biotechnology network, with extensive overseas collaboration (Dodds and Tejeda 1990). This network is illustrated in Fig. 2.1

The future research priorities for potato (Dodds and Tejeda, 1990) are:

1. Genetic mapping to locate important genes and use this information to direct the breeding programs towards specific targets.
2. Improved techniques for rapid micropropagation and preparation of virus-free material.

Fig. 2.1 International biotechnology collaboration in potato. Source: Dodds and Tejeda (1990).

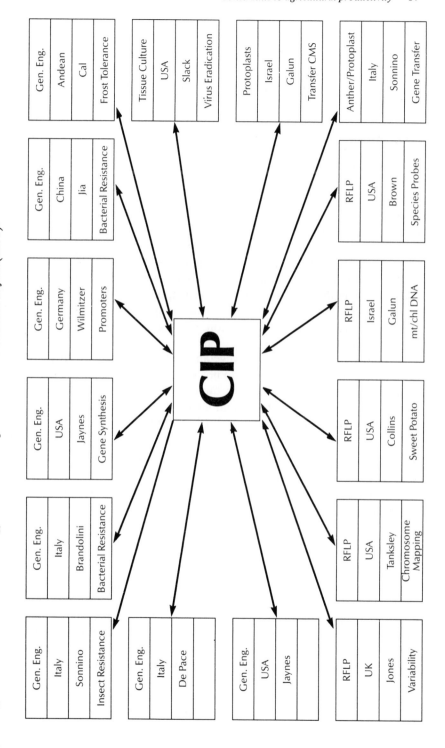

Rapeseed

Rapeseed (*Brassica* spp.) production and rapeseed oil consumption had the fastest rate of growth (9.9% per annum) of any oilseed crop over the past five years (compared with oil palm 8.5%, sunflower 5%, groundnut 3.2%, soybean 2.8%) (Scowcroft, 1990). Rapeseed is a significant crop in several Asian countries (e.g. Bangladesh, China, India, Pakistan), as well as being widely grown in North America (especially in Canada), and in several European countries. It is a temperate crop, not suited to high temperatures.

Worldwide, the crop is comprised of three related dicot species (*Brassica napus, B. campestris* and *B. juncea*). The development in Canada of high-quality, 'canola-grade' rapeseed (low erucic acid in the oil, low glucosinolates in the meal) has led to this grade becoming the international quality standard for international trade of rapeseed.

Brassica spp., particularly *Brassica napus*, are amenable to cell and tissue culture, and have been used as model systems for genetic transformation and generation. This makes the early application of new technologies to rapeseed highly likely. The ease of handling rapeseed in tissue culture, and the availability of transformation and regeneration systems to enable genetic engineering to occur, gives a major competitive advantage to rapeseed relative to other tropical vegetable oil crops such as oil palm and coconut for which such systems do not yet exist (Scowcroft, 1990).

The main research targets are quality characteristics, particularly the fatty acid composition of the oil and meal. There is a demand for speciality oils with specific fatty acid profiles for food and industrial uses.

The new technologies which are being developed for rapeseed include:

1. Genetic mapping of rapeseed, to provide information on specific traits for the breeding programs.
2. The molecular basis of biosynthesis of fatty acids, to be able to modify the fatty acid composition more precisely.

Rice

Rice (*Oryza sativa*) is the most important food crop of the developing world, and rice genetic improvement through breeding has proven to be an effective mechanism for delivering the benefits of science and technology to millions of rice farmers and consumers in the Third World. A system with international, national and regional components is now in place for producing improved rice varieties and delivering them to the farmers who need them. Biotechnology can significantly enhance the power of rice breeding programs and help produce new varieties to stabilise higher yields, improve the efficiency of production, expand the area and population base receiving benefits and further increase maximum yield potentials, thereby preventing future food shortages from occurring as demand for rice increases because of population growth and economic development.

The Rockefeller Foundation's international program on rice biotechnology is an integrated set of research, training and capacity-building activities structured to produce improved rice varieties for developing countries. The program has the objectives of:

1. Using basic science to assure the techniques of biotechnology are developed for rice.
2. Creating the capacity to conduct rice biotechnology research in rice-dependent developing countries.
3. Understanding the consequences of agricultural technical change and determining priorities for biotechnology applications.

At present, the Rockefeller Foundation commits about US$6 million/ year to support research and training at 25 institutions in industrial countries, 34 in developing countries, and two IARCs (Toennison and Herdt, 1989).

An initial assessment was made of priority needs for rice (Herdt and Riley, 1987), and support was provided to a group of the world's leading plant molecular and cell biology laboratories. This was done to move rice biotechnology research from the position of near total neglect to a level now comparable with biotechnology research on the major cereals of the industrial world. Drawing upon the knowledge base, rice biotechnology tools and training opportunities provided by the advanced laboratories, the program is now emphasising targeted research on agronomic traits, technology transfer and establishing the capability to utilise the new technologies at the international centres and in developing countries. In addition, in those developing countries having an adequate scientific establishment, the Rockefeller Foundation will help build the scientific capacity and fundamental research programs necessary for them to manage future development of rice genetic improvement technologies (Toennison and Herdt, 1989).

Progress has been rapid. A genetic map of rice based on DNA markers is now available, and through a collaborative research network the markers are being linked to genes for important agronomic traits. Protocols are available for the regeneration of rice plants from protoplasts, and for the production of fertile rice plants containing and expressing foreign genes. Several rice genes governing important traits have been cloned and characterised, as have several alien genes with potential for rice improvement. Experiments are underway to introduce these genes into rice. Scientists now predict that new rice varieties resulting from genetic engineering will reach farmers' fields by the mid-1990s, well before other cereals (Toennison, 1990).

Wheat

Wheat is the second major food of the Third World (after rice). It is particularly important in temperate regions of South Asia, China, North Africa and the Middle East. Several constraints to wheat production which are likely to benefit from the application of new technologies are: *Sclerotium* wilt; *Helminthosporium* leaf blotch; *Septoria tritici*; barley yellow dwarf virus; Karnel bunt; *Fusarium* head scab; insect pests; waterlogging; and frost damage (Larkin, 1990).

The attractive features of modern technology for wheat improvement are: genetic specificity, genetic novelty and breeding speed. The major current research requirement for wheat is a transformation and regeneration system by which new genes can be introduced and expressed (Larkin, 1990). Wheat has a large genome, approximately 17 times the size of the rice genome. This makes genetic mapping of wheat time-consuming and expensive. Several private companies in Europe have recently formed a consortium to undertake genetic mapping in wheat. CIMMYT has formed a similar consortium with Cornell University.

Findings

Constraints to agricultural productivity

There is a need to provide more opportunities for bridging courses and retraining opportunities for agricultural scientists, research managers and policymakers in NARS, IARCs and international development agencies to demonstrate the principles and potential applications of new technologies to existing research and development programs.

Greatly strengthened plant breeding programs are a necessary prerequisite for the successful adoption of modern biotechnology in developing countries.

Chapter three:
Technology assessment

Genetic engineering of crops may well prove even better suited to
the needs of the Third World ...

Introduction

Developments in biotechnology over the last decade have been accompanied by often exaggerated claims as to their likely impact on agriculture. As a result of the excitement about the scientific possibilities of genetic engineering, there was considerable loss of perspective in some quarters. Some prophets of the new technology have underestimated the pace of technical innovation, and overstated the probable flow of useful end products to agriculture. Much of the basic scientific knowledge originated in laboratories where scientists had little knowledge of the existing systems and technology of food production, into which the new technologies must fit, and with which the new technologies must compete (Cunningham, 1990). *More realistic predictions are now becoming available as to the potential applications of new technologies in agriculture and the likely time-frame for such applications.*

The most reliable estimates of the likely impact of emerging technology on agriculture are those available in the USA, where anticipated changes to the year 2000 were analysed by the Office of Technology Assessment of the US Congress (OTA, 1986). Yields of major crops are expected to increase from 0.7%/year in the case of cotton to 1.2%/year for wheat and soybeans. In the absence of the development and use of new biotechnologies in plants and animals, yield increases are expected to be between only 25 and 50% of these rates. Also, the large increases in milk production envisaged by OTA in 1986 assumed relatively rapid and widespread adoption of the use of growth hormones. This assumption may be optimistic, due to the regulatory and trade difficulties now being encountered.

Crop production

Overview

Biotechnology as applied to crop production is concerned with:

1. Agricultural microbiology, to produce microorganisms beneficial to crop agriculture.
2. Cell and tissue culture, including the rapid propagation of useful microorganisms and plant species.
3. New diagnostics based on the use of monoclonal antibodies and nucleic acid probes for the diagnosis of plant diseases and the detection of foreign chemicals, such as pesticides, in food.
4. Genetic engineering of plant species to introduce new traits.
5. New genetic mapping techniques for plant breeding programs, based on the use of restriction fragment length polymorphisms (RFLPs) as an aid to conventional plant (and animal) breeding programs.

Biotechnology includes some well-established techniques such as the use of rhizobia bacteria in nitrogen fixation of legume species; the biological control of pests; disease diagnosis based on the use of polyclonal antibodies; and plant breeding. *Support for modern biotechnology should not be at the expense of well-established technologies. It is only when there are strengths in these areas that the more sophisticated technologies of modern biology can be grafted onto existing systems (Dart, 1990a, b; Beringer, 1990).*

The two main components of modern biotechnology relevant to crop production are new diagnostics based on monoclonal antibodies and nucleic acid probes, and genetic engineering of plants. Since they currently have different potential for early application in developing countries, they are discussed separately below.

Agricultural diagnostics

New technologies are being applied to the detection and quantification of microorganisms, chemicals (particularly pesticides) and plant products. Improvements in assay technology have made possible the development of sensitive, specific, easy-to-use immunoassays for many agricultural applications. Immunoassays have been developed and are commercially available for the identification of plant pathogens, mycotoxins, pesticides and plant hormones. Monoclonal and polyclonal antibodies that have utility as specific immune reagents in immunoassays are being produced in public institutions and private companies. Nucleic acid probes have been developed for detection of all classes of plant pathogens. Widespread practical use of assays based on nucleic acid probes will be limited until stable, sensitive probe markers are developed, and not dependent on the use of radioactive labels as at present (Miller and Williams, 1990).

The major uses for modern diagnostics in agriculture are in research and development, regulation and crop management. Regulatory efforts are being assisted by the use of new diagnostic assays to detect pathogens in quarantine, mycotoxins in grains and foodstuffs, and pesticide residues. Applications of new diagnostics from crop management include: accurate disease diagnosis; selection of pathogen-free planting material; and detection and monitoring of pathogens in crops and soil. Assays suitable for on-site (farm) use are being developed for all these applications (Miller and Williams, 1990).

Monoclonal antibody production is now widely used commercially, mainly for improved diagnostics in human health care. In plant agriculture, the main use of monoclonal antibodies will be in the diagnosis of plant diseases, particularly virus diseases. Combined with ELISA (enzyme-linked immunosorbent assay, another serological technique), it is a powerful tool for detecting antigens at very low levels. It is relatively robust, much simpler than recombinant DNA technology, and would seem to be ideally suitable for promotion in developing countries. However, polyclonal antibodies can also be extremely useful, and their broader spectrum can sometimes be more useful than the highly specific monoclonal antibodies. The correct choice for the problem must be assessed on a case-by-case basis (Miller and Williams, 1990).

In considering the potential application of new agricultural diagnostics the following factors are important: (1) the technology is well developed and immediate application is feasible; (2) there is potential for application on many (probably all) crops, their pests and diseases once the specific target has been identified; (3) in some instances, the products (i.e. the diagnostics) may be more cost-effectively prepared on a contract basis in industrialised countries than in the country of use; (4) use (as distinct from preparation) of monoclonal antibodies in diagnostics does not require high technology and their application should be feasible in all countries.

Quarantine is an important area for use. Improved diagnostics would greatly increase the efficiency of any quarantine service. Because of increased efficiency and reliability of quarantine, the transfer of germplasm between countries would also be greatly faciliated. These applications would also be useful for the IARCs, in the international exchange of germplasm.

Improved diagnostics could be useful in epidemiological studies to determine the distribution and abundance of pests and pathogens. The data obtained could be used to design more effective control measures. An efficient back-up service is essential for this application in order to be able to interpret the results correctly. Improved diagnosis of diseases does not in itself control the disease. It must be supported by a system which enables effective control measures to be taken.

Developers of agricultural diagnostic products and services relevant to developing countries include private companies (currently at least 17, all of which are in industrialised countries), IARCs and public-sector institutions, often supported by international development agencies. For example, the International Potato Centre (CIP) has developed new diagnostics for identifying diseases of potato and sweet potato (Dodds and Tejeda, 1990).

The costs of being a serious participant in the development of agricultural diagnostics are high. This is a result of the technology-intensive nature of the research, development, scale-up and manufacturing efforts required. Such costs may be prohibitive to many countries. Individual countries and international development agencies can contribute to the preparation of diagnostic assays suited to needs of particular countries by co-operating with private companies and public-sector research organisations on specific aspects of kit development. These include: (1) defining end users and market needs; (2) testing prototype and final assays to assess local needs and develop interpretive data; (3) funding projects and/or guaranteeing minimum purchases of kits; (4) training end users; and (5) establishing distribution channels.

Individual countries that decide to use modern diagnostics to solve agricultural problems need to determine whether to develop the necessary systems internally, use available systems from other countries or seek partners in other countries to assist in development of specific diagnostic systems. At present, it appears more efficient for most countries to use ready-made systems and adapt them to local needs if necessary.

The possible roles of international development agencies include: (1) training in modern diagnostic technology; (2) assembling, updating and distributing information on diagnostics; (3) working with private and public sector organisations to assemble diagnostic systems appropriate for specific problems; and (4) providing funding to develop and/or implement such diagnostic systems (Miller and Williams, 1990).

Genetic engineering of plants

The key components of genetic engineering are: (1) identification and isolation of suitable genes to transfer; (2) delivery systems to introduce the desired gene into the recipient cells; (3) expression of the new genetic information in the recipient cells.

It is now theoretically possible to transfer a gene from any one organism in the biosphere to any other organism in the biosphere. The potential for improvement of crops is immense. Current research and development is concentrating on the genetic engineering of plants for herbicide resistance, insect resistance, disease resistance, improved protein composition and improved postharvest handling. Other research concentrates on the genetic engineering of insects and bacteria for improved biological control (Dart, 1990a,b; Meeusen, 1990: Whitten and Oakeshott, 1990).

It must be emphasised that there are few commercial applications of genetic engineering of plants available as yet. As agriculture in the industrial world, with all its resources, has not yet benefited greatly from genetic engineering of plants, we must not expect miracles to happen in the Third World. We are unlikely to see many major benefits to agriculture from genetic engineering of plants before the year 2000, although some early applications are likely by then. Beyond that, however, there will be increasing contributions from biotechnology to plant breeding programs, resulting in more varieties with desirable characteristics.

Using conventional plant breeding techniques, it takes about 10 years to produce a new variety. It is frequently argued that genetic engineering will dramatically reduce this time. In some cases that may be so, but it will not happen in every case. The transformation and regeneration of plants is highly variety-specific. The gene of choice will have to be introduced into a variety that can be manipulated easily in culture and then used in a conventional breeding program.

Rapid progress is being made in refining the techniques that allow a gene from one species to be transferred and expressed in another species. About 20 crops can be genetically engineered at present (Table 3.1). A current limitation to the genetic engineering of plants is the lack of efficient transformation and regeneration systems, especially for monocots, which include the world's major cereal crops. However, rapid progress is being made in the genetic manipulations of many species, and almost every month another successful plant transformation is reported.

The major limitation to the commercial development of genetic engin-eering of plants is the paucity of useful genes which, when transferred with appropriate molecular controls, will confer beneficial traits on the recipient plant.

There have been some spectacular successes in a few crops in the identification of useful genes, such as those for herbicide tolerance, insect resistance and virus resistance. These successes with crops such as tobacco and tomato have provided model systems, which now need to be adapted to more economically important plants.

Genetic engineering of crops may well prove even better suited to the needs of the Third World than to the needs of the industrial markets which are presently driving its development. Crops engineered for insect and disease resistance, for example, should offer performance to rival the use of chemical pesticides, but demand only a fraction of the capital investment to produce. They should not require sophisticated distribution systems for supply, additional equipment, specialised training, nor annual seed purchases to enable the farmer to use them. As such they appear well suited to low-input farming in the Third World (Meeusen, 1990).

Table 3.1.   Present availability of transformation and regeneration systems for crops.

Cereals	Fibre crops	Food legumes and oilseeds	Horticultural crops	Pastures	Trees
Maize Rice	Cotton Linseed (Flax)	Linseed (Flax) Rapeseed Soybean	Carrot Cauliflower Celery Chicory Cucumber Lettuce Potato Sugarbeet Tobacco Tomato	Lucerne Stylosanthus	Poplar Walnut

Genetic mapping as an aid to plant breeding

One powerful and elegantly simple outgrowth of recombinant DNA technology involves the construction of a new type of genetic map, known as an RFLP map. RFLP is the abbreviation for *restriction fragment length polymorphism,* stemming from the use of special enzymes known as restriction enzymes. These enzymes are used to treat plant (or animal) DNA so as to enable the identification of specific genetic markers at many sites throughout all the organism's chromosomes. By increasing the number and type of enzymes used, it is possible to construct a detailed map of precisely located genetic markers scattered over the entire genome of the organism.

The genetic markers in such a map are randomly distributed and far more dense than in the usual type of map of the sites of genes for morphological or biochemical traits. This magnifies enormously the power of plant (or animal) breeders to select for desired traits in otherwise conventional breeding programs.

With RFLP maps one can even conduct a breeding program without having identified the gene of interest or understanding its biochemical mechanism of action. This is done by analysing the results of a genetic cross and following how the trait is inherited with respect to the distribution of RFLP markers in the organism's chromosomes. Studying these associations makes it possible afterwards to carry out quantitative breeding programs based on qualitative data. This can substantially increase the rate of success of a breeding program.

Another powerful advantage of RFLP maps is that they make it possible to conduct breeding programs for traits controlled by more than one gene. These multigenic or polygenic traits include many characteristics of value in agriculture, e.g. drought tolerance, nutritional quality and growth rates. Using this approach one group of researchers has recently identified three genetic markers associated with water-use efficiency in tomatoes and other plants, making it possible to select for dramatic changes in water metabolism that were previously not amenable to traditional selection schemes (Martin *et al.*, 1989). This technique has great power to aid in improvements to a wide range of crop species important in the Third World, especially those for which strong conventional breeding programs already exist.

The use of plant genomic mapping is becoming more common in the USA, and was the subject of a recent conference (USDA, 1988), and review (Tanksley *et al.*, 1989). The IARCs are already making some use of RFLP techniques in their breeding programs. There is scope for much greater use of genetic mapping in the breeding of many tropical crops.

Novel means of plant virus resistance

One of the early applications of genetic engineering to tropical crops may be the introduction of novel means for plant virus resistance. In 1986, a gene encoding the coat protein of tobacco mosaic virus was introduced by transformation into tobacco and tomato plants, and conferred resistance to infection by tobacco mosaic virus. It was later demonstrated that these transgenic tomato plants had a high level of resistance under field situations (Beachy and Fauquet, 1989).

Subsequent to the first report of genetically engineered coat protein-mediated protection against tobacco mosaic virus, other research groups reported that a similar approach was applied successfully to protect plants against infection by alfalfa mosaic virus, potato virus X, cucumber mosaic virus and tobacco rattle virus. This suggests that resistance against a variety of different viruses can be achieved in plants by introducing a gene that encodes a coat protein identical to or related to that of the virus against which resistance is desired. In some way, not yet fully understood, the presence of the coat protein gene prevents that same virus infecting the plant.

In the Third World, the number of plant viral diseases is extremely high. The viruses infect, in most cases, tropical crops which are not grown in temperate regions. In many countries, few breeding programs are deployed to select plants resistant to viral diseases and the use of resistant varieties is restricted. Because of the cultural diversity and the wide variation in climatic zones, many resistant varieties are often suitable for use by only a small number of farmers. The application of coat protein-mediated resistance could be a useful approach to convert preferred crop varieties from virus-susceptible to virus-resistant, without altering other desirable qualities of the variety. It may also provide plant breeders with new sources of monogenic resistance that are easy to transfer to other plants during the normal course of plant breeding and selection.

Most of the viruses infecting tropical plants belong to the same virus groups as the temperate plant viruses, but they are either different viruses or different strains. The greatest number of these pathogens are members of the potyvirus group: for example, 13 different strains of potyviruses have been characterised in Africa infecting most of the food crops, including yam, cassava, sweet potato, maize, millet, groundnut, cowpea and other vegetables. Another important virus group, which is present primarily in the tropical countries, is the geminivirus group. These viruses are transmitted by whiteflies and leafhoppers and cause widespread, serious diseases in many different food crops, including cereals, root crops, legumes and vegetables on each of the tropical continents.

Specific biotechnology programs using genetically engineered coat protein protection should be initiated for the most economically important viruses in the Third World. To apply this type of biotechnology to a target crop, it is essential to have; (1) a basic understanding of the virus; (2) a cloned DNA representing the viral coat protein; (3) the necessary additional DNA (e.g. a promoter) to produce a functioning chimeric plant gene encoding the coat protein; (4) an appropriate gene delivery system; and (5) a tissue culture system to regenerate plants from single cells.

Although well-equipped research laboratories in many locations may be able to carry out the entire process, other research groups may need to establish collaborative relationships with laboratories presently developing this technique for plant virus resistance for temperate crops. Such collaborations may be essential to achieve resistance against virus diseases in some tropical crops for which other more classical forms of resistance are lacking (Beachy and Fauquet, 1989).

Microbial bioprocessing

Microbial bioprocessing offers significant promise to make contributions to some developing countries by the production of value added products in the food and fermentation industries (Giddings, 1990). While some countries are well positioned to exploit emerging opportunities, many will be limited in developing their own applications of microbial bioprocessing (particularly in its more sophisticated forms) by infrastructural or technical constraints. In addition to the promise microbial bioprocessing holds, however, there are the real prospects that advances in microbial bio-synthesis will lead to the production, particularly in industrialised countries, of substitutes for some agricultural exports from developing countries. Examples of this include possible substitutes for sugar, vanilla and cocoa.

Forestry

The new biotechnology offers forestry the possibilities of an array of new procedures for overcoming major constraints in woody plant improvement, protection, and utilisation (Krugman, 1990). The technical issues in forestry are different from those in agriculture. Unlike modern agriculture, which deals with a relatively few plant species – many of which have been studied and improved for years – there are numerous woody plants, often of local distribution, currently being used by rural populations in developing countries. With the exception of a few woody exotics (e.g. pines or eucalypts) most plants are endemic to a given region, and each region has its own native set of useful woody plants. These plants represent essentially wild populations and little is known about their genetic structure or even their basic reproductive biology or physiology. Most of these species have not been successfully tissue-cultured. At this time, woody plant biotech-nology is not being applied in the Third World. In spite of the constraints of limited infrastructure and little detailed knowledge of locally important species, there are still opportunities to apply the new techniques on a selective basis.

For woody plants, the new biotechnology offers the first realistic method for the identification and isolation of useful forest tree genes. Appropriate methods have been developed for woody plants, such as linkage mapping employing restriction fragment length polymorphism (RFLP), in conjunction with other techniques. Moreover, the basic methods of DNA transfer in woody plants have also been accomplished. Thus, once an appropriate system of tissue culture is established, useful DNA can be transferred. Somaclonal screening in a few woody plants has already been shown to be an efficient and rapid method for the identification of certain genotypes with desirable traits such as disease resistance. Similar methods could also be applied for the rapid identification of drought or salt tolerance. Recent advances in genetic engineering have demonstrated that natural biological control agents can be made more effective. In fact, for short-term rotation woody biomass systems, it may well be feasible to provide a degree of natural protection against both insects and certain diseases by the appro-priate gene insertion. The new biotechnology is providing an array of new methods for the early detection and identification of pathogens of woody plants, especially in seeds.

Finally, it may well be possible in the future to utilise much of the current 'waste' wood by the construction of biological wood-processing factories using genetically engineered organisms.

Livestock production

Overview

Biotechnology relevant to livestock production concerns embryo technology and many aspects of the physiology, immunology and nutrition of farm animals. The spectrum of new technologies applicable to livestock production is illustrated in Fig. 3.1. Their potential, as assessed by Cunningham (1990) and Doyle and Spradbrow (1990), is summarised below.

Some current problems and potential applications of biotechnology to their solution are listed in Table 3.2. In developing countries, cattle are the primary target for the application of new technologies to livestock production, because they are the most important species, economically and socially, and because their high individual value justifies the cost of new biotechnological inputs.

It is difficult to gauge the future pace and utility of livestock biotechnology, in both industrialised and developing countries. The experience of the past ten years has been that interesting technical developments have been achieved more rapidly than was generally predicted: for example, successful gene transfer in animals, *in vitro* fertilisation, cloning of embryos, sexing of semen are all now available (Cunningham, 1990). However, with the exception of new diagnostics, each development seems to uncover new layers of complexity in application, and as a result to push the prospect of practical applications further into the future.

A further generalisation is that the results of biotechnology are likely to be more useful to producers already using high technology than to those operating less intensively. Thus, the use of bovine growth hormone (BST) in dairy production is expected to accentuate the competitive advantage of large-scale, high-production dairy enterprises, to the disadvantage of family-scale operations. Applications of biotechnology to livestock production in developing countries need to be chosen carefully, to favour small-scale producers, and not add to already existing inequities within and between countries.

Fig. 3.1 Biotechnology in livestock production. Source: Cunningham (1990).

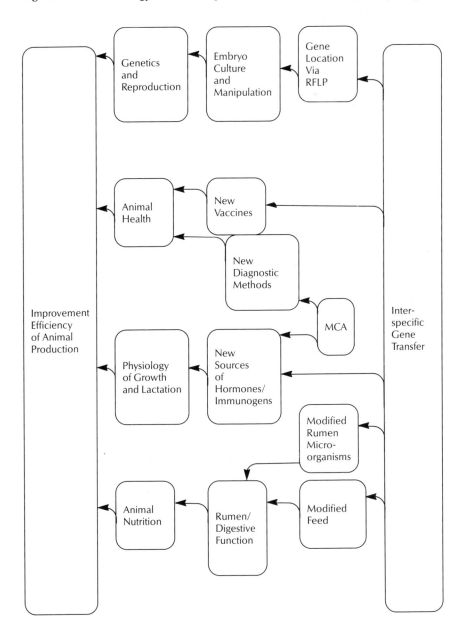

Table 3.2.   Possible applications of biotechnology to the solution of problems of livestock production in the Third World[a].

Problem	Possible solution	Scale of economic impact	Probable time to commercial use[b]
Animal, poultry, fish diseases	new vaccines and new diagnostics	large	short
Poor quality of forages	microbial treatment of forages	moderate	medium
	modification of rumen microflora	moderate/large	long
	genetic improvement of forages and their symbionts	moderate	medium
Difficulty of implementing selection programs	selection in nucleus herds, using ET, sexing	large	medium
	use of RFLP markers to assist selection[c]	moderate	medium
Difficulty of maintaining dairy cattle performance after F1 cross	user of IVF, ET and sexing[c]	large	long
Cost and environmental challenge to imported cattle	use of ET to import embyros[c]	small	short
Need for increased efficiency in intensive systems	use of rBST and rPST in dairy and pig production[c]	large	short

Notes: [a]Adapted from Cunningham (1990); Doyle and Spradbrow (1990).
[b]Short: now or <5 years; Medium: 5–10 years; Long: >10 years.
[c]ET: embyro transfer; RFLP: restriction fragment length polymorphism, or direct DNA typing of individuals; IVF: *in vitro* fertilisation; rBST: recombinant bovine or porcine somatotropin (growth hormones produced using recombinant DNA technology).

Animal disease control

Livestock diseases can be categorised in different ways, which relate both to the factors influencing a decision to control a particular disease and the development and application of suitable control measures. The term *livestock* is used here in its widest sense to include fish, poultry, domestic ruminants and exotic species. The term *disease* includes both infectious and noninfectious diseases and where necessary, genetic predispositions or resistances to certain pathogenic or metabolic disorders (Doyle and Spradbrow, 1990).

Infectious diseases can be allocated to three broad categories: (1) epidemic diseases severely damaging to livestock (e.g. trypanosomiasis, foot and mouth, Newcastle disease); (2) diseases damaging to people and livestock (e.g. rabies); (3) infectious diseases of intensive productive systems. The first two categories are of most relevance to developing countries.

Disease control in livestock has four components: (1) diagnosis; (2) treatment; (3) prevention of recurrence; and (4) eradication of disease from given populations of livestock with suitable precautions to prevent its reintroduction.

It is in these four areas where technological advances in biochemistry, immunology and molecular genetics have a role to play in improving the prospects for livestock disease control in developing countries, where all categories of infectious and noninfectious diseases represent major constraints to increased agricultural development.

There are two broad areas of technology which offer much improved

Table 3.3.   Novel animal vaccines.

Species	Disease	Producer
Avian	Coccidiosis	Genex and A.H. Robins
	Newcastle virus	Codon and Salsbury labs
Bovine	Papilloma virus	Molecular genetics
	Viral diarrhoea	California Biotechnology
	Brucellosis	Ribi ImmunoChem
	Rinderpest	USDA and University of California, Davis
Swine	Parvovirus	Applied Biotechnology
	Dysentery	Codon
Equine	Influenza	California Biotechnology
	Herpes	Applied Biotechnology
Companion	Canine parvovirus	Applied Biotechnology

Source: Cunningham (1990) and Van Brunt (1987).

means of approaching the question of diagnosis, treatment, prevention and eradication of livestock disease in developing countries: monoclonal antibody production and recombinant DNA technology. These technologies can be applied to most of the pressing problems of livestock disease control in the Third World.

The need to introduce the new technologies into developing countries, rather than simply utilise the facilities and expertise already present in the industrialised countries to make suitable vaccines, is necessitated by the fact that veterinary regulations in most industrialised countries largely preclude the importation of exotic infectious agents and livestock for research purposes.

Some novel animal vaccines, developed by the use of recombinant DNA technology, are listed in Table 3.3.

Embryo technology

Over the last 20 years, the techniques for recovery, storage and implantation of embryos have been perfected. It is now possible to superovulate cows, and to recover nonsurgically up to 30 embryos at a time. The principal benefit conferred by this technique is the ability to produce more calves from a cow than would be possible with normal reproduction. Each cow would normally give birth to about four calves in an average lifetime. With current techniques of embryo transfer, this could be increased readily to at least 25 calves. The benefits of increasing the reproductive rate of selected cows in this way are:

1. Genetically outstanding cows can contribute more to the breeding program. This is most useful if their males are being selected for use in artificial insemination programs.
2. Specially designed breeding schemes to take advantage of the increased intensity of female selection, combined with increased generation turnover, can increase (up to double) the rate of genetic change over that achievable in a conventional selection program.

Additional potential advantages from present embryo transfer technology are:

1. The possibility of increasing the twinning rate by either implanting two embryos, or by following a normal insemination with the transfer of a single embryo.
2. The rapid expansion of rare genetic stocks, for example of a new breed.
3. The transfer of specialised genotypes (e.g. pure beef breeds) of higher value into cows of lower potential.
4. Reduction of the cost of international transport of stock, by import of embryos rather than of live animals.

5. Avoiding the environmental shock to susceptible imported genotypes by having them born to dams of local breeds, rather than importing them as live animals.

The following new developments can provide additional benefits:

1. *Sexing of embryos* (now possible) or of semen (not yet possible, but being actively researched). This could increase selection intensities further, and could permit greater specialisation of the beef and milk production functions of a dual-purpose population.

2. In vitro *fertilisation.* This technique, now showing much promise in research studies, has potential for increasing the benefits achievable in breeding programs. The first commercial services based on *in vitro* fertilisation commenced in 1989 in Ireland and Britain. Its main impact will be to reduce the cost of embryos, and therefore to make embryo transfer techniques economically feasible on a wide scale, and in circumstances where the cost of present techniques cannot be justified.

3. *Embryo splitting,* to produce identical individuals. This technique has considerable application in improving the efficiency of research studies. At present levels, it has limited additional benefits in a breeding program.

4. *Embryo cloning.* The production of multiple copies of an embryo by nuclear transplantation is now possible. The main limitation is the supply of host oocytes. However, the current expectation is that repeated cloning will be a commercial reality within a few years. It could have a major impact on cattle breeding in both industrial and developing countries (Cunningham, 1990).

Genetic engineering of animals

Techniques have been developed for the insertion of DNA into animal cells in such a way that they are incorporated into the genome of the animal concerned. In animals, the usual technique is direct injection of 100–1000 copies of the DNA sequence into the pronucleus of a single embryo (Cunningham, 1990).

In 1980, the first experiment was completed which demonstrated the incorporation of exogenous genes into an animal genotype, and their subsequent expression in generations derived from the same animal. This gene transfer technology has been facilitated more in animals than in plants, because it could be superimposed on the highly developed embryo culture and transfer technology, already available for animals.

Gene transfer work in animals was pioneered in the mouse, and is now well established in that species (Jaenisch, 1988). Successful production of transgenic farm animals has so far been reported from seven centres (Table 3.4). In almost all cases the DNA transferred codes for a growth hormone.

The potential benefits of genetic engineering in animals are:

1. It is possible to insert into the animal genome the capacity to synthesise proteins which are not normal to that animal, and which may have particular value. For example, the gene for the synthesis of human blood clotting factor IX has been transferred to sheep in which it is expressed in milk (Wilmut *et al.*, 1988). The intention is that the factor, which is highly valuable as a pharmaceutical, can be isolated from sheep milk, and then be made available at greatly reduced cost. The extent to which this technology will be useful depends on the identification of proteins of high value, which are difficult to either synthesise chemically or produce biologically by conventional means.

2. It may be possible to identify genes in one species which could be transferred usefully to another species (e.g. from sheep to cattle).

3. Proteins with biological effect, such as growth hormone, can be transferred into bacterial or fungal culture and produced cheaply. They can then be administered to animals, with substantial effects on production. This is the case with growth hormone, with demonstrated large effects on milk production in cows (Chilliard, 1988) and on growth rate in pigs (Hanrahan, 1988).

4. It may be possible to locate genes with important effects (e.g. for trypanotolerance in cattle), and to use this technique to make more effective selection for the characteristic. Considerable effort is being devoted to this in Africa by the International Laboratory for Research on Animal Diseases (ILRAD).

5. It may be possible to identify marker sites throughout the genome, by genetic mapping, which can then be used to assist selection for quantitative traits, affected by many genes, such as milk production or growth rate (Chilliard and Soller, 1987).

Table 3.4.   Reported production of transgenic livestock.

Country	Centre	Species	Gene transferred[a]
USA	USDA–Beltsville	Pigs, sheep, rabbits	GH, GHRH
	Ohio State	Pigs	GH
UK	Edinburgh University	Sheep	FIX, TK, AT
Germany	Technical University, Munich	Pigs	GH
Australia	CSIRO	Sheep	GH
	University of Adelaide	Pigs	GH
Canada	University of Calgary	Cattle	?

Note: [a]GH = growth hormone; GHRH = growth hormone releasing factor;
FIX = human factor IX; TK = thymidine kinaser; AT = antitrypsin.
Source: Ward (1988).

6. Genetic mapping techniques can measure genetic distance between populations. This information could then be used to make more precise use of different breed resources in crossing and selection programs.

Hormone technology

New techniques are being developed for hormone technology. The most dramatic product of new biotechnology currently available is recombinant growth hormone, or bovine somatotropin (BST). When administered regularly to dairy cows during lactation, it increases milk output by 15–30%, and also increases efficiency of milk production. The main technical problem remaining to be resolved is that of an efficient delivery system, since current use requires regular injections (Cunningham, 1990).

The family of growth-promoting agents known as anabolic steroids has been in widespread use for many years. However, public concerns over residues in meat have led to their being banned by the European Community and by some other countries. Other metabolic agents, such as beta antagonists, have been shown in experiments to have even larger effects, but are not yet available for general use (Quirke and Schmid, 1988).

The hormone system can be manipulated to achieve improvements in productivity. One example is the use of hormones in sheep, which results in an increased ovulation rate and ultimately higher fertility. Considerable research has been undertaken to find a similar technique for use in cattle.

Aquaculture

Controlling the culture of fish and crustaceans for food probably began in Asia over 4000 years ago, but only in the last 20 or 30 years has a large commercial trade developed. Fish is a major component of the diet in many developing countries, but particularly in Asia, which dominates world production of finfish, crustaceans, molluscs and seaweed. Over 90% of production in Asia is from capture fisheries. Aquaculture is a complex process involving: seed stock production – often involving management of spawning, larvae and fingering growth; feed production – both natural and formulated: grow-out; harvesting; processing and marketing. Many of these processes rely on finely tuned skills (e.g. the fostering of parent prawns to produce spawn, an essential part of production of larvae from adults in captivity). In most aquaculture systems the same water flows freely from one farm to the next, creating problems of the spread of disease and chemical pollution between farms (Fielder *et al.*, 1990).

There has been a phenomenal annual doubling of the amount of prawns

produced in China recently, mainly through increased area under cultivation, the most rapid growth of aquaculture anywhere in the world. This is
geared mainly to the export trade, with thousands of hectares of ponds
adjacent to coastal waters. The danger with this system is that disease and
poor quality water can cause the ponds to go 'bad,' and the consequent
efflux of foul water into the coastal waters can have drastic effects on
bivalve and shellfish production. Biotechnology can help avert such
problems by providing rapid methods of disease monitoring and measure-
ment of water quality.

More accurate and more rapid biotechnological methods can be developed for disease monitoring through the use of serological techniques such
as ELISA (enzyme-linked immunosorbent assay) to identify viral, bacterial
and fungal pathogens, and through miniaturised, biochemical tests for
bacterial identification.

Feed formulation, culture of microalgae and brine shrimps for feed can
also be assisted by new analytical techniques, methods of batch culture and
storage of the feed prior to use. Aflatoxin produced by the fungus
Aspergillus flavus as a contaminant on feed materials such as corn and
groundnut, even at 20 ppb, can kill or drastically reduce the levels.

Three constraints on aquaculture need to be addressed: (1) availability
of inputs; (2) potential market for outputs; and (3) fiscal limits to
government expenditure (Fielder *et al.*, 1990).

Findings

Crop production

The likely early applications of new biotechnologies to crop production relevant to the Third World are:

1. *Agricultural diagnostics*, which could be made readily available and used widely in many countries.
2. *Genetic mapping* of major tropical crops as an aid to conventional plant breeding programs, particularly in the selection for multigenetic traits such as drought resistance and salinity tolerance.
3. *Plant virus resistance*, by genetic engineering of host plants.
4. *Novel biocontrol agents* for pest control, to reduce pesticide use.

Microbial processing

1. Novel bioprecessing techniques are likely to lead to the production of new value-added products in the food and fermentation areas.
2. The application of microbial biosynthesis in industrialised countries is also likely to lead to the production of substitutes for some agricultural commodities presently exported by developing countries.

Forestry

The likely initial applications of biotechnology in forestry are in:

1. The identification of useful forest tree genes in tree breeding.
2. The development of novel biocontrol agents against pests and diseases.

Livestock production

The likely early applications of new biotechnologies to livestock production relevant to the Third World are:

1. Embryo technology, especially for cattle.
2. New vaccines against infectious diseases.
3. New diagnostics to increase the accuracy and efficiency of disease identification.
4. Genetic mapping as an aid to livestock.breeding programs.

Aquaculture

The applications of biotechnology in aquaculture are likely to be in:

1. Rapid methods of disease monitoring by improved diagnostics.
2. Improved methods for feed formulation and storage.

Chapter four:
Socioeconomic issues

The requirement is to double food production in the Third World in
the next 25 years just to keep pace with population growth.

Introduction

The first generation of modern biotechnology has coincided with a sharp-
ened perspective on world food markets and production systems in the
industrialised world. In the last decade, the slowing down of population
growth, the levelling off in consumption patterns, and the steady progress
of agricultural technology has produced a surplus of food in the industrialised world. This has consequences both for producers and consumers.
Consumers in industrialised countries spend less than 20% of their income
on food, and are more concerned with choice, taste, health and presenta-
tion than they are with further reductions in cost of food (Cunningham,
1990).

Food production and population trends

The Third World now contains 70% of the world's population and in a
generation will contain 90%. Over recent decades world food production
has increased at a yearly rate of about 2.7%, food consumption by 2.6%
and the human population by about 2% (IFPRI, 1988). The match
between these overall figures is reassuring, but large deviations from the
average characterise particular regions and countries. In most African
countries the increase in food production has been less, and the population
increase greater; in Asia net progress in reducing hunger and poverty is not
evident despite large production gains. As a result of these inbalances, the
flow of cereals and livestock products from the surpluses in the industrialised countries to the deficit areas in developing countries has increased
rapidly. In the case of cereals, imports increased from 10 to 70 million tons
in the last 20 years (IFPRI, 1988).

Based on current population growth estimates, the requirement is to
double food production in the Third World in the next 25 years just to
keep pace with population growth. The major challenge to biotechnology,

and indeed to all food production technology, lies therefore in the developing countries, and not in the industrialised countries.

Much of the increase in world food production centres on rice, wheat and maize, the crops in which varietal improvement has been particularly significant; an intensification of production has characterised the production gains. In the livestock sector there is a similar reliance on intensification, especially with poultry and dairy production. The basic technologies of improved varieties and breeds, greater fertiliser use, extended areas of irrigation and an increase in multiple cropping have been the driving forces to the improvement in production attained in the last three decades.

In many countries the scope for improving agricultural output with existing technologies is still large. However, there are limits to what can be achieved with existing technologies when considering the need for a further doubling of food production in the next 25 years. These limits are associated with the lesser increment progressively obtained for each unit of expenditure on fertiliser and irrigation, and by the physiological barriers which constrain crop and animal yields.

Environmental considerations are also checking rapid agricultural expansion, since further declines in the area of pasture and forest land are unacceptable in many countries. *The development of new technologies, based on the application of modern biotechnology to commodities important in the Third World, will be one of the key factors necessary to obtain the substantial increases in food production required to meet expanding populations in many countries.* A widening realisation that equity and income distribution are a critical part of economic growth further modifies production strategies.

Socioeconomic impact

The time lag between investment in modern biotechnology and the production of economically useful results at present is about ten years. This time-frame is similar to that for more conventional research, in that it takes approximately ten years to produce a new crop variety, animal vaccine or pharmaceutical. The first significant production impacts from current investments in biotechnology are therefore likely to appear around the year 2000. Thereafter, modern biotechnology will become an increasingly important component of new technology for crop, forestry and livestock production.

Estimating the likely economic impact of biotechnology is difficult, due to the present lack of reliable data, the reluctance of industry to release commercially sensitive figures, the varying definitions used for biotechnology and the unpredicability of many technical developments (OECD, 1989).

Anderson and Herdt (1989) consider that the likely economic impact of modern biotechnology on agriculture in developing countries will be modest in this century, given the minimal amount of work in progress that is directly relevant to the Third World, and the likely high degree of location-specificity of some accomplishments. This is similar to the views of others predicting the impact of biotechnology on agriculture in the industrial world. Biotechnology is an unusual technology in that its scientific and social impact, including its impact on public perception (both positive and negative) have preceded its economic impact (OECD, 1989). Its economic impact is likely to increase substantially in the next century.

Biotechnology is likely to change the comparative advantages between countries and between commodities. Estimates of the potential impact of new technologies on selected commodities to the year 2000 were made by the Office of Technology Assessment of the US Congress (OTA, 1986). Lists were made of the many emerging technologies and adoption profiles developed for each. An assessment was then made of the impact of emerging technologies for several major crops and animals. For example, yields of major crops are expected to increase from 0.7%/year for cotton to 1.2%/year for wheat and soybean. In the absence of the development and use of new technologies in plants and animals, yield increases are expected to be only 25–50% of the estimated rates. The estimated rates of yield increases for most crops and animals are about the same, or even less than the annual yield increases obtained for major commodities over the past 30 years (OTA, 1986).

Barker (1990), in assessing the potential economic impact of biotechnology in the Third World, considers that biotechnology will contribute to modest but continued increases in productivity of the major crops of the order of 1.5–2.2%/year. These continued productivity increases are unlikely in the absence of modern biotechnology, due to yield plateaux being reached for some of the major crops.

The message for the Third World is clear: not to apply new technologies for the improvement of their food and export commodities will put them at a competitive disadvantage in the international marketplace.

Biotechnology is potentially a 'scale neutral' technology, but some applications could contribute to serious, long-term negative consequences for global trade and development. To the extent that biotechnology will speed up the innovation process, it will tend to accentuate the impact of price distortions on trade and development. Reduction of protectionism in the industrialised countries, removal of policies that penalise agriculture in the developing countries, and reduction of the Third World debt burden are seen as critical steps in enhancing the productivity impacts and mitigating the negative equity impact of biotechnology (Barker, 1990).

The role of the private sector in biotechnology will continue to grow. It is likely that the private sector will develop those technologies and deal with those countries where the profits seem assured. The private sector is likely to be willing to do business with perhaps ten developing countries where public-sector research and infrastructure are already well established. This would leave out at least 50 other countries, including most of Africa. These countries need initially to strengthen more traditional applied biological science and plant breeding in order to develop the capacity to utilise the products of biotechnology. The international development agencies need to ask what steps can be taken to protect public-sector interests and encourage the development of biotechnologies beneficial to the broad range of developing countries.

Barker (1990) makes several recommendations concerning:

1. Policy reforms, including: (a) reduced protectionism in the industrial world, and (b) reduced Third World debt burdens;
2. Strengthening global agricultural research, through: (a) stronger links between public research agencies in industrialised and developing countries, (b) support for improved educational and research institutions in developing countries, and (c) targeted donor support for training in biotechnology.
3. Assessing the potential and socioeconomic impact of biotechnology through (a) work by multi-disciplinary treams of biological and social scientists on determining priorities for biotechnology investment, (b) examining returns to investments in public biotechnology, and (c) similarly, the social returns to private biotechnology investments.

Buttel (1990) in his assessment of the likely social impact of biotechnology, noted that biotechnology has real potential for the Third World in the long term. Most biotechnology will be directed at the perceived most financially profitable opportunities. These only coincidentally, and probably quite unlikely, will be of direct relevance and utility to resource-poor farmers in developing countries, unless there is redirective intervention by concerned bodies such as international development agencies.

The dominance of the private sector of industrialised countries in modern biotechnology creates some difficulties for information-sharing with developing countries, and potential problems of political support within OECD (Organization for Economic Cooperation and Development) countries for technical assistance grants by bilateral development agencies.

Biotechnology may contribute to restructuring of markets for food commodities of food-feedstock commodities for new bioprocessing and to the replacement of markets for traditional industrial/beverage crops through development of new substitutes. International development agencies should be aware of such possibilities for their commodity forecasting work and investment activities.

The growing private-sector participation in R and D, combined with often stagnating public R and D activity, provides conflicting stresses, yet also provides opportunities and a need for new policy pathways for international development agencies. Biotechnology should not be seen as a substitute for investment in more conventional agricultural research. Rather it is a new aspect which will be complementary to other approaches to agricultural and economic development.

Trade impact

At present, competition in biotechnology is principally among OECD countries. For OECD countries, being 'competitive' in biotechnology means preparing for the future by reinforcing the scientific, technical, staffing, industrial and information infrastructure required for biotechnology (OECD, 1989).

One of the potential areas of negative impact from biotechnology on developing countries is the development of new methods of producing commodities (or their industrially important constituent chemicals). There may be the possibility of production of the desired compound in tissue culture rather than by traditional means of field production (e.g. cocoa butter from tissue culture). Alternatively, there may be ways to alter the quality of a temperate crop so that it is able to substitute for a previously imported tropical commodity (e.g. rapeseed could substitute for imported coconut oil, if its oil composition was changed to include a higher concentration of lauric acids, the components of coconut oil which gives it a price premium for industrial use).

The trade substitution effects of biotechnology were apparent even before the widespread interest in genetic engineering. This is part of an old pattern, where industry captures markets away from agriculture (OECD, 1989; Goodman *et al.*, 1983). Two areas of real or potential negative impact of biotechnology on developing country agriculture are: (1) enzyme-based sweeteners to reduce the need for cane sugar imports; and (2) *in vitro* plant propagation and cell tissue culture of various species.

In the mid 1960s, it became technically feasible to obtain sweeteners from any starch by the use of new enzyme techniques. These developments allowed the production of large volumes of high fructose corn syrup from maize in the USA. This replaced much of the imported cane sugar from the Philippines and elsewhere. Although this is not an effect of modern biotechnology, it is illustrative of a real negative impact of traditional biotechnology on Third World exports.

The potential impact of tissue culture production of biological products is more speculative. Several private companies are actively working on different possibilities for the displacement of plant-derived products presently grown in the Third World. These include flavours such as cocoa butter, pharmaceuticals such as codeine and digitalis, and essences such as jasmin (Kenny and Buttel, 1985). At present, the production costs in tissue culture are high, and it is only economic for high-value products worth more than US$700/kg. A lowering of production costs would make tissue culture more competitive for a wider range of products (OECD, 1989).

Conclusion

A common consequence of change in technology is larger farms displacing smaller farms, and centres of production moving to new areas that have developed a greater comparative advantage. A particular problem for the small-scale farmer is that technology adoption has often been initiated by packaging together several techniques which complement each other. Improved grain varieties are often bred to respond to larger amounts of fertiliser, to critically timed cultivation, and to require specific disease and pest control measures. The improvement of dairy cattle is usually associated with genetic change via the introduction of artificial insemination, production recording, progeny testing, and more intensive feeding and husbandry practices. Extending technology packages to small farms often proves more difficult and expensive than for large farms, and technology change can disadvantage the poorer farmers.

Technological change provides a dilemma, particularly for smaller countries and poorer farmers. Without a change in technology the balance between the supply and demand for basic food commodities in many countries is likely to be inadequate; with a changing technology smaller farmers have greater difficulty, and need more time and resources to adjust to the new production circumstances. More and better grain varieties, fertiliser and irrigation will continue to provide the basis of the production response needed. As the constraints of energy prices, environmental deterioration and production plateaux become more binding, new innovations will be sought increasingly through the application of biotechnology.

The potential for positive impact of biotechnology on the Third World will be enhanced by the formation of strategic alliances between national governments, international development agencies, and private companies presently investing in biotechnology. This is particularly relevant to those countries where public-sector research and infrastructure are already well developed, since these are where private companies are likely to increase their activities in the future. *The challenge for international development agencies lies especially with the many other countries which will require additional public-sector investments to develop their capacity to adapt and utilise the products of biotechnology.*

International development agencies need to take special steps to encourage the development and use of biotechnologies beneficial to the broad range of countries in Africa, Asia and Latin America. This will require substantial public-sector investments in educational institutions to upgrade the teaching of biological sciences, and continued support for research institutes to strengthen more traditional biological research (including plant breeding), as a necessary prerequisite for the use of biotechnology. Support for policy reforms in world trade, reduction in protectionist policies and in Third World debt burdens will also be important.

The likely socioeconomic effects of biotechnology are positive in terms of increasing the productivity of tropical commodities, opening up new opportunities for the use of marginal lands, and reducing use of agro-chemicals. They are also potentially negative, in that they offer the possibility of producing high-value products in tissue culture in industrialised countries, and thus displacing crops presently grown for export in the Third World. The potential negative substitution effects should be monitored by international development agencies, and strategic adjustments made where feasible. These actions could include loans or grants to assist in crop diversification in individual countries.

Findings

There is a need for continuing socioeconomic assessment of the likely impact of biotechnology, especially in relation to:

1. Determination of priorities for investments in biotechnology, particularly those of likely benefit to resource-poor farmers and to smaller countries.

2. Development of an 'early warning system,' concerned with the identification of potential negative effects from the substitution of tropical commodities by novel products produced in industrialised countries and determination of the strategic adjustments possible to minimise these effects on individual countries.

3. Identification of social and economic returns to investment in public- and private-sector biotechnology.

Chapter five:
Biotechnology in industrialised countries

The major reason for the greatly increased role of the private sector
... is that ... the process and/or the product is protectable.

Introduction

Modern biotechnology was first used commercially in the mid 1970s in the USA, when several new companies were established to develop and market biological products resulting from the use of technologies, such as recombinant DNA technology, monoclonal antibodies and novel bioprocessing techniques. The first of the new biotechnology companies established was Genentech in 1976. It produces Inulin (a form of insulin), the first pharmaceutical sold commercially, the production of which was based on novel genetic techniques. Over 600 new biotechnology firms have been established, mainly in the USA. The early products marketed have almost all being pharmaceuticals and diagnostics. These products are being developed primarily for markets in industrialised countries. Sales of monoclonal antibodies for human health care amounted to several hundred million dollars in 1987 (OECD, 1989). Few novel products for agriculture are yet on the market.

Market predictions

The total world market for agrichemicals and seeds in 1987 was estimated to be US$113 billion in farm-level sales. Of this, 51% (US$59 billion) was for fertilisers, 29% (US$32.6 billion) for traded seeds, and 20% (US$22.6 billion) for pesticides. Estimates on future markets for biotechnology products vary widely, ranging from US$10 billion to US$100 billion, for farm-level sales by the year 2000 (OECD, 1989). Some industry sources suggest that a realistic estimate would be US$10 billion for total world sales at the farm level of novel products in agriculture by the year 2000. Of this seeds would comprise approximately US$7 billion, agricultural microbiological products US$1 billion and veterinary products US$2 billion.

The emphasis is likely to be on novel products taking an increasing share of existing markets with some restructuring of those markets, rather than any major expansion in the markets for agriculture-related products.

Research and development investments

The major change in the funding of agricultural research in industrialised countries in the past decade has been the substantially increased role of the private sector, largely in funding research in modern biotechnology. In 1985 it was estimated that approximately US$4 billion was spent on research and development activities in modern biotechnology worldwide (Table 5.1). Of this, US$2.7 billion (67%) came from the private sector. These trends have continued, with total industrial research and development expenditure in the OECD countries in 1987 estimated to be in the range of US$3–4 billion (OTA, 1988a; OECD, 1989).

Table 5.1. Geographic distribution of R and D expenditure (in US$ millions) on modern biotechnology (1985 estimate).

	Private sector	Public sector
USA	1500	600
EEC	700	300
Japan	400	200
Others	100	200
Total (worldwide)	4000	

Of the US$4 billion spent on modern biotechnology in 1985, it was estimated that US$900 million (22%) was spent on agriculture-related research, and US$3 billion spent on other forms of modern biotechnology, mainly for human health care (Table 5.2).

Within the total estimated agricultural biotechnology expenditure of US$900 million in 1985, US$600 million (66%) was spent on seeds and US$300 million (34%) on agricultural microbiology (Table 5.3). An estimated US$550 million (60%) of the total expenditure on agricultural R and D came from the private sector (Table 5.3).

There are also substantial public-sector investments in biotechnology in industrialised countries. The public sector research investments in the USA were critical to the initial findings in recombinant DNA technology. The development of monoclonal antibodies originated in a public-sector institution in the UK. The 1988 OTA report on US investments in biotechnology suggests that public sector investments in the USA have increased

Table 5.2. Sector distribution of R and D expenditure (in US$ millions) on modern biotechnology worldwide (1985 estimate).

	Agricultural biotechnology	Other
Private	550	2150
Public	350	950
Total (worldwide)	4000	

substantially in the past three years (OTA, 1988a). Recent OECD reports also suggest that most OECD countries are increasing their public-sector spending in biotechnology, including in the agricultural sector (OECD, 1988, 1989).

The major reason for the greatly increased role of the private sector in modern biotechnology is that, for many of the new technologies, the process and/or the product is protectable. A company is able to appropriate many of the benefits of its research investments in modern biotechnology. Private companies are therefore much more willing to invest in modern biotechnology than in more traditional agricultural research for which it is more difficult to obtain proprietary rights to the technology generated. The exceptions are agrichemicals and hybrid seeds, where there have been substantial private sector R and D investments, since both these types of products are able to be protected.

The companies investing in modern biotechnology in agriculture are: (1) new biotechnology firms; (2) major agrichemical and seed companies; and (3) other major companies, particularly food companies. Several invest at least US$5 million per year on R and D in modern biotechnology. In 1986 there were approximately 134 companies in the USA involved in agricultural biotechnology, of which 70 were new biotechnology firms (OTA, 1986). The commercial application of biotechnology is presently dominated by American and European companies, with increasing interest from Japanese companies.

Table 5.3. R and D expenditure (in US$ millions) on agriculture-related biotechnology worldwide (1985 estimate).

	Private sector	Public sector
Agricultural microbiology	200	100
Seeds	350	250
Total (worldwide)	900	

Table 5.4.   Biotechnology companies and sector of interest.

Source	USA	Japan	France	UK	Federal Republic of Germany
Agriculture	73	12	5	15	2
Chemical	37	31	1	4	4
Diagnostics	141	15	3	10	6
Food	18	17	2	12	1
Pharmaceuticals	65	28	2	9	4
Veterinary	54	2	3	6	0
Total	388	105	16	56	17

Source: OECD (1988).

The number of companies investing in biotechnology in the USA, Japan, France, UK and the Federal Republic of Germany are listed in Table 5.4. The figures indicate the relatively large number of specialist biotechnology companies in the USA and, to a lesser extent, in the UK, and the diversification of many Japanese companies into biotechnology. In contrast, Germany and France have generated few new companies, but this is not necessarily a measure of commitment nor potential success. Increased resources from a few large companies may be more effective than establishing many small new companies (OECD, 1988).

Several large agrichemical companies are developing their markets for new products by the purchase of seed companies. This allows them to integrate modern biotechnology into conventional plant breeding programs (Fishlock, 1989) *There are now a series of mergers and acquisitions in train as research and development investments are rationalised in the light of more realistic market forecasts and better estimates of the likely time-frames for the availability of novel products.*

New biotechnology firms

New biotechnology firms are defined as entrepreneurial ventures established after 1976 'whose sole functional is research, development and production, using biotechnological means' (OTA, 1984).

Many of the new biotechnology firms were established by scientists from the public sector who realised the commercial potential of their work. *Biotechnology is said to be 'the first business with enough glamour to persuade eminent scientists that the entrepreneurial spirit and academic respectability are not mutually exclusive'* (Wyke, 1988). The companies were often set up close to public-sector institutions, with encouragement and sometimes partial financial support from government. Many of the new biotechnology companies in the USA are clustered close to those universities with a strong tradition of research in the biological sciences.

The characteristics of the new biotechnology firms are that they are usually innovative, conduct 'leading edge' research, and are knowledgeable about basic biological research being conducted at public-sector institutions. The firms often have R and D contracts with leading scientists at public-sector institutions, with proprietary rights to new technologies generated. The average size of American biotechnology companies in 1984 was 15 PhD scientists per firm (OTA, 1984).

The new biotechnology firms had several sources of funds. These included: (1) venture capital; (2) equity; (3) research contracts; (4) joint venture and partnership arrangements; (5) profit from product sales; and (6) public sources of funds. The main source used by new firms to become established was venture capital. Increasingly, the new biotechnology firms have been undertaking contract research for major companies, and entering into licensing arrangements under which larger companies market the novel products developed.

The main advantage of the new companies is in research. Their small size, flexibility and entrepreneurial climate provides a favourable environment to attract high-quality scientists, capable of the scientific innovation necessary for new product development.

The major disadvantages of the new biotechnology firms are that they are often undercapitalised and have cash flow problems; and that they do not have marketing and distribution networks and brand name recognition to facilitate the sale of novel products internationally.

The new biotechnology firms are likely to concentrate on niche markets, particularly those in which the major transnational companies are not interested; or to enter into joint ventures and licensing agreements with major companies for the marketing of novel products developed by the new biotechnology companies.

These partnerships between the new biotechnology firms and the major companies have given the USA an initial competitive advantage internationally, in the commercialisation of biotechnology. This competitive edge may not be sustained as the R and D (Research and Development) investments made by the major European and Japanese companies since 1980 begin to yield results.

Major companies

Investment strategies

The major players in modern biotechnology in industrialised countries are the transnational pharmaceutical, agrichemical, seed and food companies in USA, Europe and Japan. The comparative advantages of these companies relative to the new biotechnology firms are that: (1) they have sufficient cash flow from other sources to invest in high-risk research for several years before realising a return on their investment; this level of investment is beyond the scope of the new biotechnology firms; (2) they have international marketing networks through which to sell new products; and (3) they are familiar with the regulatory procedures governing the release of new products, which is helpful in understanding the regulatory procedures required for the release of novel products.

The major companies are investing in modern biotechnology in three ways: (1) in-house research to develop novel products for which the company has proprietary rights, a brand name and an international distribution and sales system; (2) collaborative ventures with scientists in public sector institutions, especially universities, with specific contract research being undertaken in the public-sector institution, under a mutually agreed, confidential sharing of information and negotiated proprietary rights to new processes and products; and (3) joint ventures with new biotechnology firms, either by licensing arrangements to market novel products previously developed by a new biotechnology firm, or by the contracting of research for product development to a new biotechnology firm, under a mutually agreed sharing of results, proprietary rights and profits.

The Japanese government, through its Ministry of International Trade and Industry (MITI), has identified the commercialisation of biotechnology as one of its key targets for the future of Japanese economic development in the next century. Japanese companies did not begin investing in modern biotechnology until 1980. Few new biotechnology firms have been established, and the commercialisation of modern biotechnology is being undertaken largely by established Japanese companies.

Research in Japan is being undertaken in the private sector (approximately 61%), public sector organisation (13%) and universities (26%) (OTA, 1984). MITI and associated ministeries such as the Ministry of Agriculture have established R and D 'clubs' to bring together public and private-sector institutions with expertise in the different technologies required for the successful commercialisation of biotechnology. The Ministry of Agriculture has established two clubs to foster the application of modern biotechnology to plant breeding and agricultural microbiology, respectively. Japan's main comparative advantage in biotechnology is in bioprocessing, based on its existing industrial fermentation industries for amino acid production and other biological products.

Crop production

Most of the transnational agrichemical companies in the USA and Europe have taken the decision to invest in plant biotechnology. Some have large in-house research programs; some are investing substantially in new biotechnology firms, without developing an in-house capacity; others are contracting research with universities and other public-sector bodies at modest levels, and observing R and D results emanating from other major companies. Several transnational companies spend at least US$5 million/ year on modern biotechnology. Precise figures on R and D investments are difficult to obtain, for commercial reasons.

Several major agrichemical companies are developing their potential markets for plant biotechnology products by the purchase of seed companies. Ownership of seed companies provides a well-established marketing and distribution system, in addition to a germplasm base developed over many years of conventional plant breeding. This provides the basic material which can be genetically engineered and reintroduced into conventional breeding programs to produce new varieties, with novel characteristics. *The most successful players will be those who are able to integrate new biotechnologies into conventional plant breeding programs.*

It is the strategy of most agrichemical companies that the sale of novel varieties will compensate them for the loss in revenue due to reduced use of agricultural chemicals (a worldwide trend). In the case of herbicide-tolerant varieties, a proprietary package of seeds and herbicides may be marketed.

Similarly, several transnational companies are investing in agricultural microbiology, with the view to developing novel biological control agents to replace partially the use of pesticides, fungicides and herbicides. In many industrialised countries the environmental lobby is campaigning for the reduced use of agrichemicals. There is also a growing public perception that foods produced with minimal chemical application are desirable products, which command a price premium. (Private-sector strategies are discussed more fully in a recent OECD report (OECD, 1989).)

Livestock production

The applications of modern biotechnology to animal production are concerned primarily with: (1) the prevention and control of animal diseases by the production of vaccines and diagnostic aids; (2) growth promotion; and (3) genetic improvement of breeds.

The important features of the commercialisation of novel animal health products are that: (1) recombinant DNA technology used to make human vaccines is applicable to the production of many animal vaccines; (2) monoclonal antibody and nucleic acid probe technologies developed for use with human pathogens are often applicable to the development of diagnostics for animal diseases; (3) the markets for many animal health products are small, and new biotechnology firms may be able to compete with larger companies for these markets; and (4) the regulatory requirements for the commercial introduction of veterinary vaccines are simpler than for similar products for human health care.

Animal biotechnology is of particular interest to many new biotechnology firms. A number of companies working on animal biotechnology are also involved in human health care, and their animal research is a spin-off from their medical research. Several new biotechnology firms have been active in the development of diagnostic products, growth promotants and vaccines for animals. The interaction between human and animal biotechnology has meant that novel animal-related products have been developed ahead of plant-related products.

The companies involved in the commercialisation of biotechnology in animal agriculture are primarily the major American and European pharmaceutical and chemical companies and new biotechnology firms, mainly in the USA. Several major companies are involved in the production of animal growth promotants, some in collaboration with new biotechnology firms in the USA. The main European efforts in animal health are in the development of vaccines against rabies and foot-and-mouth disease. Japanese companies are little involved in animal-related biotechnology as yet.

Conclusion

A new feature of modern biotechnology has been public/private sector collaboration, with private companies often contracting specific research projects at public-sector institutions and, in return, having proprietary rights to the technology generated.

A new pattern of agricultural research funding is emerging from the linkages of biotechnology to both universities and private firms, with increased research financing coming from sources beyond traditional government appropriations. In several industrialised countries, many public research institutions are now required to raise a substantial part of their research budget from nongovernment sources via contractual research, licensing agreements and royalties. A new pattern of agricultural research activity is the result, one in which traditional methods of open interaction and communication in research are being rapidly altered; and where the potential availability of intellectual property rights for new technologies influences the extent of private investment.

Findings

1. Continued public-sector investments in biotechnology and creative partnerships between public and private sector interests are critical for a country in establishing a competitive strategy in biotechnology.

2. National governments and international development agencies need to devise new mechanisms, including innovative funding arrangements, to encourage the greater participation by private-sector companies in biotechnology in developing countries, both by local private-sector firms and transnational companies. These would facilitate the acquisition and adaptation of new technologies to the needs of Third World agriculture.

Chapter six:
Biotechnology in developing countries

How much to invest ... on what commodities to invest, and what
staffing and infrastructure are required ...

Introduction

Countries are at different stages in determining their policies in relation to
biotechnology, and in putting these policies into practice. The countries fall
into three main categories:

1. Countries with interest but no direct involvement in biotechnology as
yet.
2. Countries that have a national biotechnology policy and program, most
of which is concerned with traditional biotechnology. They have established collaborative links with industrialised countries for the training of
scientists, and for the acquisition of new technologies to adapt to local
problems.
3. Countries which have a national biotechnology policy and program,
including an in-country program in modern biotechnology, in addition to
collaborative linkages overseas, in both the public and private sectors.

There are at present low barriers to entry into modern biotechnology.
The barriers to entry are likely to become higher, because: (1) economies
of scale will emerge, especially in bioprocessing to allow the large-scale,
cheap manufacture of new products; (2) complementarity with other technologies, such as information technology and microelectronics, will become
more important as processes become automated; and (3) access to
large-scale marketing and distribution networks will become increasingly
important in the sale of novel products. These three factors tend to favour
the large companies in industrialised countries.

There are, however, other factors which favour entry by developing
countries into biotechnology. These are: (1) the large and growing
domestic markets in developing countries; (2) the opportunities to identify
niche markets in which the major transnational companies are neither
active nor interested; and (3) the availability of plant genetic resources for
many of the world's major agricultural crops, most of which originated in
the Third World. The wild species and land races contain many useful

genes, some of which could be transferred by means of genetic engineering to cultivated varieties of crop plants to produce useful commercial varieties. Entry into biotechnology is not restricted to large countries, if small countries select appropriate market niches, in a manner analogous to new biotechnology firms.

The potential application of modern biotechnology to agriculture in developing countries has been termed a 'Biorevolution' by some and compared to the 'Green Revolution' with its release of high-yielding wheat and rice varieties by the IARCs and many NARSs (Kenny and Buttel, 1985).

There are several significant institutional differences between the 'Green Revolution' and the 'Biorevolution' (Table 6.1). The differences stem

Table 6.1.   Characteristics of the 'Green Revolution' and the 'Biorevolution'.

Characteristic	Green Revolution	Biorevolution
Crops affected	Wheat, rice	Potentially all crops, including vegetables, fruits, export commodities and specialty crops
Other products affected	None	Animal products Pharmaceuticals Processed food products Energy
Areas affected	Some developing countries, mainly better locations	Potentially all countries and all areas, including marginal lands
Technology development and dissemination	Largely public or quasi-public sector	Substantial private sector involvement
Proprietary considerations	Patents and plant variety protection not important	Many processes and products patentable and protectable
Capital costs of research	Relatively low	Relatively high
Research skills required	Conventional plant breeding and other agricultural sciences	Molecular and cell biology expertise plus conventional plant breeding skills and expertise in other agricultural sciences
Crops displaced	None, but traditional varieties and land races replaced by high-yielding varieties	Potentially many

Source: Modified from Kenny and Buttel (1985).

primarily from the large private-sector investments in R and D in modern biotechnology in the industrialised countries, and the proprietary nature of many of the new products and processes.

National agricultural research systems

Modern biotechnology will have a profound impact upon agricultural research and agricultural productivity. It will be a principal source of technological change in agriculture in the coming decades. Like all major technological changes, biotechnology will bring about distributional impacts among different classes of producers, among exporters and importers, and among producers and consumers. The developing countries, whether they like it or not, will find themselves in the wake of these waves of change and, therefore, need to organise their NARSs accordingly (Javier, 1990).

Two major features associated with modern biotechnology have significant implications for the way agricultural research will be organised and managed in the future, namely: (1) the increasing participation of the private sector in agricultural research and development; and (2) the shortening of the lag phase between discovery and application, and the narrowing of the gap between basic science and applied research. Both trends have been under way for some years in the industrialised countries in the science-based industries. Modern biotechnology is accelerating the process for agriculture.

Expertise in the basic sciences is in short supply in many countries. Biological scientists are usually located in institutions with no current responsibility for agricultural research. *Mobilisation of the basic biological scientists usually found in the universities to solve agricultural problems will require new policy and institutional arrangements and additional financial resources in most countries.*

Many countries need to upgrade their teaching of science by incorporating modern biotechnology into the teaching of the traditional agricultural and biological sciences. For those countries with more advanced systems of higher education, there is a need to enhance the capacity of the universities to provide training in molecular biology, genetics, biochemistry, physiology, immunology and chemical engineering (Holloway, 1990).

The increasing dominance of the private sector in agricultural biotechnology R and D in the industrialised countries, and their increasing partnership with the universities, may increase the cost of access by individual countries to advances in science and technology, which previously have been freely available as public goods. Countries may have to accustom themselves to the idea of negotiating and importing not only finished agricultural technology, but also research technology, through commercial channels. Under these circumstances, the IARCs may assume greater importance, as windows of access to advanced technologies for the NARSs, especially for smaller countries. It is likely that the larger and more technologically advanced countries will negotiate their own bilateral arrangements on access to new technologies. The IARCs may be able to advise them, if requested, on the cost and suitability of certain technologies being considered for import.

Private-sector investment in agricultural R and D is rising in many countries in Asia and Latin America. These private-sector investments are important particularly in internationally traded commodities, where countries are most vulnerable to competition. Policies on patents, tax incentives, contract research, mutual sharing of resources, and the appropriate division of labour between public and private research entities need to be developed in most countries to facilitate greater private-sector participation in agricultural biotechnology (Javier, 1990).

Biotechnology includes a wide array of products, processes and research areas. The key management problem in the immediate term, given the chronic scarcity of the resources, is deciding which applications in agriculture should be given priority. *The key questions to consider are: how much to invest in biotechnology relative to other agricultural R and D; on what commodities to invest; and what staffing and infrastructure are required to support these investments?* Countries vary widely in their needs and scientific potential. However, even for those that now have the capacity to engage in recombinant DNA research, the most obvious early choices are those biotechnological applications which are commercially applied elsewhere (such as improved diagnostics). A framework within which to make choices of priority research areas in biotechnology would be extremely valuable. An initial framework has been developed for rice (Herdt and Riley, 1987). A similar system may be valuable for other commodities.

A common organisational dilemma is whether to establish a central, separate modern biotechnology institute or to graft biotechnology onto programs at existing agricultural research institutes. Since expertise, facilities and funds are limited, it is expedient in many countries to think in terms of a central laboratory where a critical mass of resources and effort can be mobilised. There are disadvantages to the centralisation of staff and resources for biotechnology, since the integration of biotechnology into existing agricultural research will be essential for its successful application. Whichever institutional mechanism is chosen, it is critical to ensure that there are strong collaborative linkages between the new cell and molecular biologists and the agricultural scientists located in other institutes in the country. A central laboratory can then service the needs of several institutes without the necessity of duplicating equipment and facilities at every institute with an interest in biotechnology.

Modern biotechnology will not diminish the need for conventional agricultural research. Thus, the demand for new expertise and companion facilities will exert additional pressure on the already limited resources allocated to agricultural research. Support for biotechnology development by national governments will build upon the pattern of assistance to agricultural research they already provide. Changes in this pattern also seem inevitable, including a greater proportion of equity funding to facilitate risk sharing; the encouragement of partnership arrangements between private industry and public sector organisations; an increased involvement in supporting training in biology; particular attention to the assessment of research priorities; and encouragement of local investments in agricultural biotechnology.

Selected country experiences

The status of biotechnology was assessed initially in ten countries selected to provide a sample of countries of different sizes and at different stages of development in relation to biotechnology. The countries in which the initial surveys were undertaken were Brazil, Brunei, China, India, Indonesia, Malaysia, Mexico, Philippines, Singapore and Thailand (Anon, 1989a).

Each country study was conducted by consultants (primarily from the country itself), who arranged visits to principal institutions and laboratories and interviews with policymakers responsible for biotechnology and senior personnel engaged in biotechnology research, development and commercialisation. Contacts were made with both public and private sector institutions in each country.

Completion of a standard questionnaire facilitated the generation of a uniform set of core data from each country which allowed the data to be summarised for all countries and trends identified. This also facilitated the input of the various country reports into a computerised data base which is being assembled for subsequent use. The general findings of the ten-country case studies are presented below. They are supplemented with more detailed reports for each country, which are available from ISNAR.

The assessment of factors that contribute to the operation of effective national biotechnology programs are summarised in Table 6.2 for the ten countries surveyed. Nine countries viewed biotechnology as a high priority area, warranting increased investments of the national R and D budget in science and technology. The exception was the relatively oil-rich country of Brunei which has only a small program in biotechnology, as part of an

The percentage of gross domestic product (GDP) devoted to research and development in science and technology is shown in Table 6.2. The data show that, in general, the more advanced and more industrialised the country, the higher the percentage of GDP it invests in science and

Table 6.2. Assessment of factors that contribute to the operation of effective national biotechnology programs in ten selected countries.

Country	% GDP devoted to science and technology	Public-sector budget biotechnology (US$ millions)	Well-defined national policy	Domestic programme adequate	Domestic budget adequate	Human resources adequate	Private-sector activity	Dedicated biotech companies	Venture capital	Adequate patent protection	Adequate environmental regulations	University industry co-operation	International technology transfer
Brazil	0.6	7–15	1	2	3	1	2	3	3	3	3	2	1
Brunei	0.1	NA	3	3	1	3	3	3	3	NA	NA	3	3
China	NA	NA	1	2	2	2	3	3	3	3	3	3	2
India	0.9	31+	1	3	1	1	2	3	3	3	3	2	1
Indonesia	0.3	NA	1	2	2	3	3	3	3	NA	NA	NA	3
Malaysia	NA	0.4	1	2	2	NA	3	3	3	NA	NA	NA	NA
Mexico	0.6	3	1	3	3	2	3	3	3	3	3	3	2
Philippines	0.2	1	1	2	3	1	1	3	3	3	3	2	2
Singapore	0.9	10+	1	3	1	1	3	3	2	3	3	2	1
Thailand	0.3	7–14	1	1	2	1	3	3	3	3	2	3	2

Notes: NA – information not available except for the first two columns; 1 – favourable; 2 – acceptable; 3 – low.
Source: Anon (1989a).

technology. These are also the countries that have chosen to place more emphasis on biotechnology. The percentage of GDP devoted to science and technology ranges from 0.1 to 0.9 and the annual public-sector budget in biotechnology from US$0.5 to 30 million.

Although the size of the country is obviously a factor that determines the absolute level of the domestic budget for biotechnology, there is a general relationship between state of economic development and commitment to science and technology and investments in biotechnology.

There was a general recognition in the countries surveyed that biotechnology will make a significant contribution to increased agricultural productivity and production. Given the extent to which the countries are dependent on agriculture, the high priority assigned to food security and the recognition of the need to maintain international competitive advantage, particularly for exports in the international markets, it is not surprising that high priority was assigned to agricultural biotechnology within the context of the long-term national development plans.

All countries, with the exception of Brunei, have a strong commitment at the policy level to consign significant and additional resources to biotechnology. However, some countries have not been able to implement these policies effectively because of the absence of a well-defined national biotechnology program or because of financial constraints on the domestic budget.

A few countries have well-formulated policies and national programs and are fortunate to have access to reasonable amounts of financial resources in the domestic budget to support a national program in biotechnology. Most other countries, although committed at a policy level, have not, for diverse reasons, been in a position to elaborate a well-defined and focused national biotechnology program. These same countries have also suffered from domestic budget constraints, some because of significant national debts and others because of lack of economic stability and growth (Table 6.2).

The development of a national biotechnology strategy and a plan of activities would be helpful in the implementation of an effective national biotechnology program, which makes efficient use of domestic and external resources.

The public sector is the major supporter of biotechnology activities in all countries surveyed. Typically, because of the multidisciplinary nature of biotechnology, the Ministry of Science and Technology assigns a particular institute responsibility to co-ordinate the activities of government institutions and universities.

In some countries, for example Singapore, the major activities are concentrated in a central laboratory. However, it is more common for a network of several centres of excellence in biotechnology to be established at existing agricultural and biomedical research centres.

The availability of adequate domestic funds to support biotechnology activities varies greatly between countries, with India and Singapore having adequate levels of funding; China, Malaysia, Thailand and Indonesia with acceptable but not favourable levels of financial support; and Brazil, Mexico and the Philippines suffering from a severe shortage of funds (Table 6.2).

In considering the composition of agricultural biotechnology programs, the predominant activity is plant biotechnology, followed by industrial applications mainly featuring fermentation, with animal biotechnology receiving least attention. Within plant biotechnology, micropropagation and tissue culture is the predominant activity with emphasis on high-value crops such as ornamentals. There is much interest in the production of pathogen-free material, particularly for the vegetatively propagated crops, and some interest in germplasm conservation and the production of secondary metabolites.

Crop improvement usually features the techniques of more traditional biotechnology such as anther culture and embryo rescue rather than the utilisation of more advanced techniques such as genetic mapping as an aid for plant breeding programs. New diagnostics for plant diseases, based on monoclonal antibodies, are also attracting wide interest.

Basic research studies and genetic engineering are restricted to a few countries, and usually to one or two laboratories within those countries. There are some laboratories in the countries surveyed that have a capacity to conduct basic research and genetic engineering of the same quality as the centres of excellence in the industrialised countries. These laboratories represent an important institutional and national capacity which can act as centres of excellence and provide a solid foundation for strengthening the capacity of national and regional programs in biotechnology.

Industrial applications of agriculture concentrate on fermentation processes, including the traditional foods, brewing and processing of agricultural waste. Waste disposal is an important priority for countries such as Malaysia with large amounts of waste biomass from palm oil processing for disposal. Production of single-cell protein is a longer-term objective which is of interest to some countries. In animal biotechnology, embryo transfer, growth hormones and diagnostics and vaccines are the main areas of interest.

Of the countries surveyed, one-half had sufficient trained staff to conduct current programs. Staffing could quickly become a constraint if more financial resources were devoted to biotechnology or if there was a change in emphasis in favour of modern biotechnology. Some countries, including India, Singapore, Brazil, Thailand and the Philippines had adequate staffing, with Mexico and China having acceptable levels but no reserve capacity. Indonesia has a shortage of staff and this is likely to be exacerbated when planned projects become operational.

A common trend across the countries is the low level of activity of the private sector. The data in Table 6.2 show that with the exception of Singapore, and to a lesser extent India and Brazil, the private sector does not have a viable presence in biotechnology in the ten countries surveyed. This is a notable contrast with the industrialised countries where the private sector is a major investor in biotechnology, with transnational companies and new biotechnology firms playing complementary roles.

In the countries surveyed, the formation of new biotechnology companies has not taken place (with the exception of a few companies in Brazil; Table 6.2). There are several reasons for this. Biotechnology is considered to be a long-term, high-risk venture which does not lend itself to the operations of the smaller, indigenous, private-sector companies in the developing world which are more geared to short-term, production-oriented objectives and cannot support long-term R and D objectives. Secondly, the movement from universities of academic entrepreneurs who founded new biotechnology firms in the USA has not usually occurred elsewhere. Finally, the venture capital that is required to establish the new companies is not currently available in any of the countries surveyed.

Most countries surveyed reported that there were already in place many incentive schemes in the form of easier credit, tax shelters or write-offs, which are designed to stimulate local companies, new biotechnology firms, and transnational companies to make more investments in biotechnology. However, in most cases the desired effect to increase the involvement of industry is not being realised.

The lack of adequate patent protection, which is common for all countries surveyed, is certainly a factor (but not the only one) which dissuades transnational companies from conducting research activities on marketing biotechnology products in certain countries. The local subsidiaries of the transnational companies tend to rely on their parent companies to conduct R and D, on the basis that opportunities for product commercialisation for local markets can be explored later when markets can be more easily assessed.

The lack of involvement of the private sector has important implications. First, industry usually has a comparative advantage in finishing, commercialising and distributing agricultural products. Lack of industry involvement represents a missing link in the production chain in relation to product commercialisation and distribution (Bollinger, 1990; James and Persley, 1990; Raff, 1990).

Secondly, the low level of activity by industry does not allow a national biotechnology program to be implemented which benefits fully from the complementary inputs from the public and private sectors, both of which represent national assets. More specifically, the lack of industry–university co-operation programs (see Table 6.2) is a serious disadvantage to the country and to the institutions. Experience in the industrialised countries would indicate that the development of such co-operative programs is important in the transfer of technology between the public and private sectors.

The creation in individual countries of a common forum through biotechnology societies and government-sponsored symposia designed to identify national goals and objectives, and the corresponding respective and complementary roles of the public and private sectors, may help resolve some of the current constraints to industry participation.

Patent protection is certainly a factor which dominates the decision-making of transnational companies in relation to the marketing of bio-technology products. It is also a controversial issue for several developing countries. The reluctance to support patent protection relates to concerns that it would suppress in-country development of biotechnology, and lead to dominance by transnational companies; and a fear that local genetic resources would be modified and patented by others, thereby denying the country access to the products or, alternatively, limiting access through purchase at unacceptable costs. As a result of these concerns, all surveyed countries currently do not provide patent protection for plant and animal products (Table 6.2). Some countries, such as Mexico, have made conces-sions to allow future patent protection for biotechnology processes but still excluding protection for plant and animal products (patent issues are explored further in Chapter 8).

Similarly, with the exception of Thailand, none of the countries surveyed yet have guidelines and regulations governing the controlled experimen-tation and release of novel products into the environment. Given the increasing international concern about the environment and the sustain-ability of agricultural systems, the absence of appropriate regulatory procedures could be a significant impediment to the timely introduction of agricultural biotechnology applications to the developing countries. This concern is recognised in most countries. Assistance from industrialised countries and international development agencies could greatly facilitate the preparation of suitable national guidelines to regulate the release of novel organisms into the environment (regulatory issues are discussed further in Chapter 7).

International transfer of biotechnology is well under way in most countries. It could be considerably strengthened through escalating the level of activity via international development agencies and bilateral agreements. This is particularly important for countries such as Brazil, Mexico and the Philippines which have demonstrated political willingness to devote more resources to biotechnology, but cannot do so because of severe financial constraints.

To date, the international transfer of technology has been limited to the transfer from the public sector in the industrialised countries to the public sector in the developing countries. Given that the private sector invests at least as much as the public sector in biotechnology on a global basis, it is likely that private companies in the industrial world will have appropriate technology suitable for transfer and adaptation in the Third World.

This represents a new challenge to the international development agencies, the developing countries and IARCs. There is a need to seek new mechanisms, including innovative funding arrangements, to facilitate the transfer and adaptation of technologies from the private sector in the industrialised countries to both the public and private sectors in the Third World. Possible strategies for this transfer are discussed in Chapters 9 and 11, and James and Persley (1990).

Findings

The development of a *national biotechnology strategy*, involving both the public and private sectors is important for implementation of an effective national biotechnology program that makes efficient use of both domestic and external resources.

Chapter seven:
Regulatory issues and environmental release

Action has been taken in several countries to ... ensure
environmental safety and public health in biotechnology applications
in agriculture.

Introduction

A major issue that will affect the application of biotechnology to agriculture is the regulatory climate governing the release of novel products. A safe and efficient regulatory environment is in itself a comparative advantage in biotechnology.

The regulatory requirements come from the legitimate need to assure environmental safety and public health. The issue is being considered by many national governments, as well as several international agencies including the OECD (1986) and the European Commission. The scientific community is also addressing the issues involved in the release of genetically engineered microorganisms (Sussman *et al.*, 1988).

Many OECD countries have existing regulations concerning the release of new biological products from any source. These usually provide sufficient protection for the approval of the commercial sale of the products of biotechnology, since existing legislation governing the release of products such as agrichemicals, biological control agents, animal vaccines and new plant varieties is as relevant to the products produced with the aid of modern biotechnology as it is to more conventionally produced products.

In countries without existing statutory laws or guidelines governing the release of agrichemicals and biological products the position is less satisfactory. Several United Nations organisations are reviewing the needs of their member countries to ensure that safe and efficient regulatory processes are available to all countries. The concerned agencies include the Food and Agriculture Organization (FAO), the World Health Organization (WHO), the United Nations Environment Programme (UNEP), the United Nations Industrial Development Organization (UNIDO) and the International Labour Office (ILO). The international scientific community via the International Council of Scientific Unions (ICSU) and the Rockefeller

Foundation are also providing useful advice (Barton, 1989).

The new requirements are for guidelines to cover the handling of genetically engineered organisms at the experimental stage, and methods for risk assessment prior to widespread commercial use. These early releases were the subject of well-publicised concerns about the consequences of their deliberate release, especially in the US and Europe. However, much experience has been gained in the past several years that will be of relevance to developing countries in the establishment of national guidelines to govern the release of genetically engineered organisms. By mid 1990, there have been 187 deliberate releases into the environment of genetically engineered plants and microorganisms in 17 countries: USA (93); France (28); Canada (18); Belgium (12); UK (8); The Netherlands (6); New Zealand (5); Australia (4); Spain (3); Denmark (2); Italy (2) and one each in Germany, Sweden, Finland, Argentina, Mexico and Japan. None of these releases have led to any demonstrated adverse environmental effects.

Ever since human beings have lived in settlements, grown crops and herded animals, we have selected organisms with desired properties and bred them with the intention of producing offspring that also possessed those properties. This has been a slow process, carried out over thousands of years. Although the process of domestication has changed greatly the plants and animals that now dominate the planet, we have not in the past regulated these activities by law (Millis, 1990).

Over the past 100 years knowledge from basic science has enlarged our understanding of genetics and provided powerful new tools which can be applied to breeding programs. The discovery and use of mutagenic agents allowed mutation rates to be greatly accelerated in certain species and thus provide a larger pool of changed genes from which to select desired forms. This was exploited by the fermentation industry, especially during the 1940s, and played a major role in the evolution of valuable industries for the production of antibiotics, organic acids, enzymes and amino acids. The manufacture of these products using live mutant strains is conducted in many countries under codes of good manufacturing practice and general safety (Millis, 1990).

The codes are directed towards ensuring that the products are safe for industrial use or human consumption, and that those engaged in their production have safe working environments. It is widely agreed that the mutated microbial strains used for production are beneficial to the community and are not hazardous to workers or the environment. It is also widely agreed that no special surveillance is required simply because mutant strains are used.

The advent of recombinant DNA technology in the early 1970s led to the unprecedented step (at the Asilomar Conference) of scientists formally examining the possible consequences of their work and placing constraints on some avenues of research until the implications of such work could be thoroughly considered. This action was taken more in awe of the potential of new techniques for manipulating biological materials than out of any clear understanding of dangers that might arise from such manipulations. It is instructive to note that as new research results and risk assessment information were accrued, the constraints on research were refined, and many were relaxed or eliminated. Attitudes in several countries towards the commercial use of genetically engineered organisms in the environment is following a similar evolution (Millis, 1990).

Regulatory approaches

Substantial relevant action has been taken in several countries to explore the measures necessary to ensure environmental safety and public health in biotechnology applications in agriculture. Four definitive statements on regulating environmental applications of genetically engineered organisms have been published in the USA: These came from: (1) the US National Academy of Sciences (NAS, 1987); (2) the Office of Technology Assessment of the United States Congress (OTA, 1988b); (3) the Ecological Society of America (Tiedje *et al.*, 1989) and (4) the National Academy of Sciences – National Research Council (NRC, 1989). They share common judgements on the costs and benefits of the new technologies and in the appropriate approaches for regulatory authorities to take. All are cautiously optimistic that the benefits of the use of new technologies outweigh the risks, and that mechanisms exist to allow regulatory authorities to assess the instances of low, medium and high levels of risk, and to enforce appropriate safeguards. The UK Royal Commission on Environmental Pollution has also reported recently on approaches to be taken in the UK (Anon, 1989b).

The 1987 US National Academy of Science study states:

1. There is no evidence that unique hazards exist either in the use of recombinant DNA techniques or in the movement of genes between unrelated organisms.

2. The risks associated with the introduction of genetically engineered organisms are the same in kind as those associated with the introduction of unmodified organisms and organisms modified by other methods.

3. Assessment of the risks of introducing genetically engineered organisms into the environment should be based on the nature of the organism and the environment into which it is introduced, not on the method by which it was produced.

4. Recombinant DNA techniques provide a powerful and safe new means for the modification of organisms.

5. *Genetically modified organisms will contribute substantially to improved health care, agricultural efficiency and the amelioration of many pressing environmental problems that have resulted from the extensive reliance on chemicals in both agriculture and industry.*

6. *The timely development and the rational introduction of recombinant DNA-modified organisms into the environment depend on the formulation of sound regulatory policy that stimulates innovation without compromising good environmental management.*

7. The scientific community urgently needs to provide guidance to both investigators and regulators in evaluating planned introductions of modified organisms from an ecological perspective.

The OTA study (OTA, 1988b) deals in greater detail with genetic and ecological issues, and with the problems of risk assessment and the impact of public perception. Its major conclusions are:

1. While there are reasons to be cautious, there is no cause for alarm at the prospect of environmental applications of genetically engineered organisms.

2. Adequate prerelease safety review of planned introductions is now possible, even though some scientific uncertainties remain that will be resolved only with practical experience.

3. None of the small-scale field tests proposed within the next several years are likely to result in an environmental problem that would be widespread or difficult to control.

4. In many cases realistic small-scale field tests are likely to be the only way potential risks from commercial-scale uses of genetically engineered organisms can be evaluated.

5. In evaluating the potential risks associated with these new technologies, the appropriate question is not 'How can we reduce the potential risks to zero?' but 'What are the relative risks of the new technologies compared with the risk of the technologies with which they will compete?' Furthermore, 'What are the risks posed by overregulating, or failing to develop fully the new technologies?'.

6. Because the critical issues differ from application to application, a flexible review process, founded in critical scientific evaluation and adaptable to the requirements of particular cases, can serve industry and the public interest well without being unduly burdensome.

7. Although there are enough uncertainties that introductions should be approached with caution, a large body of reassuring data, derived chiefly from agriculture, supports the conclusion that with the appropriate regulatory oversight, the field tests and introductions planned or probable in the near future are not likely to result in serious ecological problems.

8. It should be possible to sort planned introductions into broad categories for which low, medium or high levels of review are appropriate. A review procedure that involves the flexible, adaptable, case-by-case review of proposed planned introductions deemed to involve significant risk is most prudent at present.

The recent report from the Ecological Society of America (Tiedje *et al.*, 1989) is the most focused on ecological issues. Chief among its conclusions are:

1. Ecological oversight of planned introductions should be directed at promoting effectiveness while guarding against potential problems. The diversity of organisms that will be modified, functions that will be engineered and environments that will receive altered organisms makes ecological risk evaluation complex. While the complete exemption of specific organisms or traits from regulatory oversight is not recommended, the Society supports the development of methods for scaling the level of oversight needed for individual cases according to objective, scientific criteria, with a goal of minimising unnecessary regulatory burdens.
2. Genetically engineered organisms should be evaluated and regulated according to their biological properties (phenotypes), rather than according to the genetic techniques used to produce them.
3. Evaluating the benefits and risks of biotechnology products requires expertise in many scientific disciplines ... For society to realise the full benefits of biotechnology, interdisciplinary research and graduate training programs are needed to expand the expertise of the scientific community at large.

Guidelines for environmental release

The OECD (1986) has published a study on the commercial use of genetically engineered organisms in industry, agriculture and the environment with the long-term objective of harmonisation of biosafety policies in OECD countries. The formal recommendations of the OECD's Council in 1986, concerning safety considerations for applications of recombinant DNA organisms in industry, agriculture and the environment, are given in Table 7.1. This statement contains the only presently available, internationally endorsed recommendations on guidelines for the release of genetically engineered organisms. It formed the basis for the national guidelines developed in several OECD countries when adapted to suit national needs. It could also provide a useful starting point for other countries presently engaged in establishing national guidelines. The terms of reference of the national genetic manipulation advisory committee in Australia (which are consistent with the OECD guidelines) are given in Table 7.2.

The need for statutory regulations on the conditions governing release

Table 7.1.   Recommendation of OECD's council concerning safety considerations for applications of recombinant DNA organisms in industry, agriculture and the environment.

The Council,

Considering that recombinant DNA techniques have opened up new and promising possibilities in a wide range of applications and can be expected to bring considerable benefits to mankind;

Recognising, in particular, the contribution of these techniques to improvement of human health and that the extent of this contribution is expected to increase significantly in the near future;

Considering that a common understanding of the safety issues raised by recombinant DNA techniques will provide the basis for taking initial steps towards international consensus, the protection of health and the environment, the promotion of international commerce and the reduction of national barriers to trade in the field of biotechnology;

Considering that the vast majority of industrial recombinant DNA large-scale applications will use organisms of intrinsically low risk which warrant only minimal containment consistent with good industrial large-scale practice (GILSP);

Considering that the technology of physical containment is well known to industry and has successfully been used to contain pathogenic organisms for many years;

Recognising that, when it is necessary to use recombinant DNA organisms of higher risk, additional criteria for risk assessment can be identified and that these organisms can also be handled safely under appropriate physical and/or biological containment;

Considering the assessment of potential risks of recombinant DNA organisms for environmental or agricultural applications is less developed than the assessment of potential risks for industrial applications;

Recognising that assessment of potential risk to the environment of environmental and agricultural applications of recombinant DNA organisms should be approached with reference to, and in accordance with, information held in the existing data base, gained from the extensive use of traditionally modified organisms in agriculture and the environment generally, and that with step-by-step assessment during the research and development process potential risk should be minimised;

Considering the present state of scientific knowledge;

Recognising that the development of general international guidelines governing agricultural and environmental applications of recombinant DNA organisms is considered premature at this time;

Recognising that there is no scientific basis for specific legislation to regulate the use of recombinant DNA organisms;

On the proposal of the Committee for Scientific and Technological Policy:

1. RECOMMENDS that Member countries,

(a) share, as freely as possible, information on principles or guidelines for national regulations, on developments in risk analysis and on practical experience in risk management with a view to facilitating harmonisation of approaches to recombinant DNA techniques;

(b) examine their existing oversight and review mechanisms to ensure that adequate review and control of the implementation of recombinant DNA techniques and applications can be achieved while avoiding any undue burdens that may hamper technological developments in this field;

Table 7.1. cont.

(c) recognise, when aiming at international harmonisation, that any approach to implementing guidelines should not impede future developments in recombinant DNA techniques;

(d) examine at both national and international levels further developments such as testing methods, equipment design and knowledge of microbial taxonomy to facilitate data exchange and minimise trade barriers between countries. Due account should be taken of ongoing work on standards within international organisations, e.g. WHO, CEC, ISO, FAO, MSDN[a];

(e) make special efforts to improve public understanding of the various aspects of recombinant DNA techniques;

(f) watch the development of recombinant DNA techniques for applications in industry, agriculture and the environment, while recognising that for certain industrial applications, and for environmental and agricultural applications of recombinant DNA organisms, some countries may wish to have a notification scheme;

(g) ensure that assessment and review procedures protect intellectual property and confidentiality interests in applications of recombinant DNA, recognising the need for innovation while still ensuring that all necessary information is made available to assess safety.

2. RECOMMENDS, with specific reference to industrial applications, that Member countries:

(a) ensure, in large-scale industrial applications of recombinant DNA techniques, that organisms which are of intrinsically low risk are used wherever possible, and handled under the conditions of Good Industrial Large-Scale Practice (GILSP) described in the report;

(b) ensure that, when a risk assessment using the criteria defined in the report indicates that a recombinant DNA organism cannot be handled merely by GILSP, appropriate containment measures, in addition to GILSP, and corresponding to the risk assessment are applied;

(c) encourage, in large-scale industrial applications requiring physical containment, further research to improve techniques for monitoring and controlling non-intentional release of recombinant DNA organisms.

3. RECOMMENDS, with specific reference to agricultural and environmental applications, that Member countries:

(a) use the existing considerable data on the environmental and human health effects of living organisms to guide risk assessments;

(b) ensure that recombinant DNA organisms are evaluated for potential risk, prior to applications in agriculture and the environment by means of an independent review of potential risks on a case-by-case basis[b];

(c) conduct the development of recombinant DNA organisms for agricultural or environmental applications in a stepwise fashion, moving, where appropriate, from the laboratory to the growth chamber and greenhouse, to limited field testing and finally, to large-scale field testing;

(d) encourage further research to improve the prediction, evaluation, and monitoring of the outcome of applications of recombinant DNA organisms.

Table 7.1. cont.

---

4. INSTRUCTS the Committee for Scientific and Technological Policy to:

(a) review the experience of Member countries in implementing the principles contained in the report;

(b) review actions taken by Member countries in pursuance of this Recommendation and to report thereon to the Council;

(c) consult with other appropriate Committees of the OECD in developing proposals for a co-ordinated future work programme in biotechnology.

Notes: [a]World Health Organization (WHO); Commission of the European Communities (CEC); International Standards Organisation (ISO); Food and Agriculture Organisation (FAO); Microbial Strains Data Network (MSDN).

[b]Case-by-case means an individual review of a proposal against assessment criteria which are relevant to the particular proposal; this is not intended to imply that every case will require review by a national or other authority since various classes of proposals may be excluded.

Source: OECD (1986).

are not embraced enthusiastically when they are excessively rigid, expensive and time-consuming. Such regulations may be self-defeating in that they encourage unauthorised experimentation. Guidelines established by a technical committee rather than legislative regulations are the preferred choice in some countries as they have the advantage over statutory regulations of flexibility and rapid change in a field where the expansion of knowledge is rapid (Millis, 1990).

In implementing appropriate guidelines, the key step is the formation of a hierarchy of biosafety committees, at the institutional and national level, that carry tiered responsibility for:

1. Certifying the security of the facilities used for genetic engineering work according to the category of risk.

2. Approving proposals at the institutional level for work of the lowest category of risk and forwarding to the national committee for review any proposals where the institutional committee is uncertain of the risk, or where the genetic construct is one of perceptibly high risk.

3. Ensuring that advice from the national biosafety committee is followed, including the appropriate training of workers.

4. Inspecting containment facilities regularly.

5. Providing, at regular intervals, the national biosafety committee with a list of all current work involving genetic manipulation (World Bank, 1989).

Table 7.2. Terms of reference for a genetic manipulation advisory committee.

Having regard to the Government's wish for a voluntary monitoring system for recombinant DNA technology and for advice to be provided to the Minister on the continuing assessment of the hazards associated with the production and/or application of material incorporating recombinant-DNA molecules unlikely to occur in nature, *the Committee shall*:

1. Establish and review as necessary Guidelines for both physical and biological containment and/or control procedures appropriate to the level of assessed risk involved in relevant research, development and application activities.
2. Review relevant proposals, except those that relate to research performed under laboratory containment conditions, and recommend any conditions under which this work should be carried out, or that the work not be undertaken.
3. Consult with relevant Government agencies and other organisations as appropriate.
4. Report to the Minister at least annually. Report promptly to the Minister on breaches of the Guidelines referred to in (1) above, and on other relevant matters referred to the RDMC by the Minister.
5. Establish contact and maintain liaison with such monitoring bodies in other countries and with international organisations, as is appropriate.
6. As necessary, advise on the training of personnel with regard to safety procedures.
7. Collect and disseminate information relevant to the above having due regard to the special circumstances relating to proprietary information.
8. Oversee the work of a Scientific Sub-Committee whose terms of reference, in addition to (3), (5), (6) and (7) above in so far as they relate to laboratory contained research, are set out below.

The Scientific Sub-Committee shall:

1. Enter into discussions directly with scientists and the institutions where they work, and with fund granting bodies in determining the conditions under which laboratory contained research with recombinant DNA molecules should be carried out.
2. Review proposals for such research and recommend any conditions under which experiments should be carried out, or that work not be undertaken.
3. Provide technical advice to the Committee and contribute to the function of the Committee in relation to laboratory contained research.

The Committee shall report within five years of its establishment on the need for continuing to monitor recombinant DNA activities.

Source: RDMC (1987).

Risk assessment criteria

Tiedje *et al.* (1989) in their study for the Ecological Society of America present a list of the criteria important to consider with respect to the safety of planned introductions. It develops further risk assessment criteria established in Australia, and shows how such criteria might be linked together in a flexible review scheme that should be of great interest and value to regulators.

The US National Academy of Science (NAS) and its National Research Council (NRC) has recently released a report on the establishment of a framework for decisions on the introduction of genetically modified microorganisms and plants into the environment (NRC, 1989). The study was charged with identifying criteria for defining risk categories and recommending ways to assess the potential risks associated with introducing modified organisms into the environment.

The report reiterates the principle of the earlier Academy document (NAS, 1987) that *safety assessment of a recombinant DNA-modified organism 'should be based on the nature of the organism and the environment into which it will be introduced, not on the method by which it was modified'.* The principle that evaluation should be of the product and not the process by which the product is obtained is reemphasised. It also points out that although genetic modification by molecular methods may be more powerful and capable of producing a wider range of phenotypes, *'no conceptual distinction exists between genetic modification of plants and microorganisms by classical methods or by molecular methods that modify DNA and transfer genes'.*

The 1989 NRC report also concludes that investigators modifying microorganisms for environmental introduction should assess the influence of genetic alteration on the organism's phenotype and the mobility of the altered trait. It is highly unlikely that moving one or a few genes from a known pathogen to an unrelated nonpathogen will confer pathogenicity on the recipient. If the recipient is itself a pathogen, increased virulence for particular hosts may result. If modifications of this latter type are contemplated, special attention must be paid to them.

The NRC also notes that in some cases persistence is not desirable and uncertainty exists about the microorganism's effect on the immediate environment. When assessing risk in these cases, the most relevant phenotypic properties relate to the persistence of the microorganism and its genetic modification. Evaluation of phenotypic properties raises questions about the fitness of the genetically modified microorganism, the tolerance of the introduced microorganism to physicochemical stresses, its competitiveness, the range of substrates available to it and, if applicable, the pathogenicity, virulence and host range of the introduced microorganism.

The NRC also recognises that there is a long history of utility and safety in the use of plants and microorganisms. *'Society has benefited greatly from the use of genetically modified microorganisms and plants, and field testing is essential to increase our knowledge about the relative safety or risk of large-scale use of genetically modified organisms and to determine the potential utility of the modified organisms'.*

With regard to the *field-testing of genetically modified plants*, the NRC report concludes that:

1. Plants modified by classical genetic methods are judged safe for field testing on the basis of experience with hundreds of millions of genotypes field-tested over decades. The current means for making decisions about the introductions of classically bred plants are entirely appropriate and no additional oversight is needed or suggested.

2. Crops modified by molecular and cellular methods should pose risks no different from those modified by classical genetic methods for similar traits. As the molecular methods are more specific, users of these methods will be more certain about the traits they introduce into the plants. Traits that are unfamiliar in the specific plant will require careful evaluation in small-scale field tests where plants exhibiting undesirable phenotypes can be destroyed.

3. At this time, the potential for enhanced weediness is the major environmental risk perceived for introductions of genetically modified plants. The likelihood of enhanced weediness is low for genetically modified, highly domesticated crop plants, on the basis of our knowledge of their morphology, reproductive systems, growth requirements and unsuitability for self-perpetuation without human intervention.

4. Confinement is the primary condition for ensuring safety of field introductions of classically modified plants.

5. Depending on the crop species, proven confinement options include biological, chemical, physical, spatial, environmental and temporal isolation, as well as size of field plot.

6. Plants grown within field confinement for experimental purposes rarely, if ever, escape to cause problems in the natural ecosystem.

7. Established confinement options are as applicable to field introductions of plants modified by molecular and cellular methods as to introductions of plants by classical genetic methods.

With regard to the *field testing of genetically modified microorganisms*, the NRC report concludes that:

1. The precision of many of the molecular methods allows scientists to make genetic modifications in microbial strains that can be fully characterised, in some cases to the determination of specific alterations of bases in the DNA nucleotide sequence.

2. The molecular methods have great power because they enable scientists to isolate genes and to transfer them across biological barriers.

3. Although field experience provides considerable information about some microorganisms (e.g. rhizobia, mycorrhizae and many plant pathogens and biocontrol agents), information regarding the ecology of microorganisms and experience with planned environmental introductions of genetically modified microorganisms is limited compared with that regarding plants. However, no adverse effects have developed from introductions of genetically modified microorganisms. Ecological uncertainties can be addressed scientifically with respect to genetic and phenotypic characterisation of the microorganisms as well as by consideration of environmental attributes such as nutrient availability. Field tests of genetically modified organisms can go forward when sufficient information exists to permit evaluation of the relative safety of the test.

4. The likelihood of possible adverse effects can be minimised or eliminated by appropriate measures to confine the introduced microorganism to the target environment, for example, by introducing 'suicide' genes, as they become practicable, into the organisms.

Fig. 7.1 Framework to assess field testing of genetically modified plants. Souce: Modified from NRC (1989).

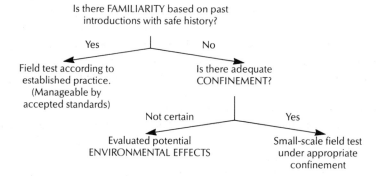

Fig. 7.2 Familiarity tests for genetically modified plants. Source: Adapted from NRC (1989).

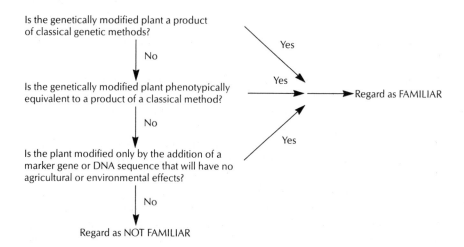

The NRC report also provides frameworks for the evaluation of risk for the release of plants and microorganisms (Figs 7.1 – 7.9). These frameworks are based on the following criteria:

1. Are we familiar with the properties of the organism and the environment into which it may be introduced?
2. Can we confine or control the organism effectively?
3. What are the probable effects on the environment should the introduced organism or a genetic trait persist longer than intended or spread to nontarget environments?

When the familiarity standard for a plant or microorganism has been satisfied such that reasonable assurance exists that the organism and the other conditions of an introduction are essentially similar to known introductions, and when these have proven to present negligible risk, the introduction is assumed to be suitable for field testing according to established practice.

The familiarity criterion is central to the suggested framework of evaluation. Its use permits decision makers to draw on past experience with the introduction of plants and microorganisms into the environment, and it provides future flexibility. As field tests are performed, information will continue to accumulate about the organisms, their phenotypic expression and their interactions with the environment. Eventually, entire classes of introductions may become familiar enough to require minimal oversight.

'Familiar' does not necessarily mean safe. Rather, to be familiar with the elements of an introduction means to have enough information to be able to judge the introduction's safety or risk.

Fig. 7.3 Confinement tests for genetically modified plants. Source: Modified from NRC (1989).

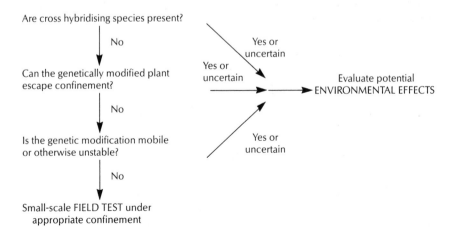

When knowledge of the type of modification, the species being modified or the target environment is insufficient to meet the familiarity criteria, the proposed introduction must be evaluated with respect to the ability to confine or control the introduced organism and to the potential effects of a failure to confine or control it. The results of the latter evaluations will define the relative safety or risk of a proposed introduction (NRC, 1989).

The framework for microorganisms and plants differ in nomenclature

Fig. 7.4   Potential environmental effects. Appropriate questions for specific applications to be added by users of the framework for the release of genetically modified plants. Source: Modified from NRC (1989).

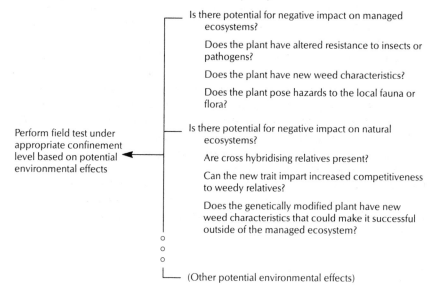

Fig. 7.5   Framework to assess field testing of genetically modified microorganisms. Source: Modified from NRC (1989).

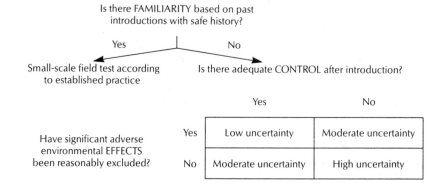

and in emphasis on particular issues, mainly because of differences in life cycles, mechanisms of gene transfer, dispersal and containment or control procedures, persistence and environmental factors. Fewer proposed field tests of microorganisms than plants may meet the familiarity criteria because the data base, from a history of planned introductions, is more limited. Means to confine plants are well established and can be relatively simple, whereas means to control microorganisms are more difficult (NRC, 1989).

The NRC evaluation of the scientific issues and their proposed frame-

Fig. 7.6 Familiarity tests for genetically modified microorganisms. Source: Adapted from NRC (1989).

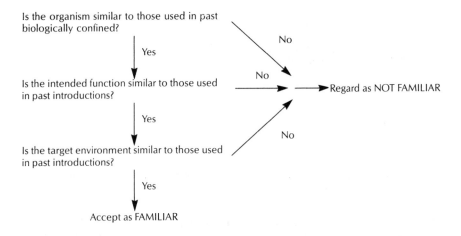

Fig. 7.7 Control. Appropriate questions for specific applications to be added by users of the framework for the release of genetically modified microorganisms. Source: Modified from NRC (1989).

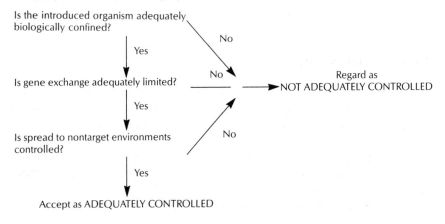

Fig. 7.8   Biological confinement. Appropriate questions for specific applications to be added by users of the framework for the release of genetically modified microorganisms. Source: Modified from NRC (1989).

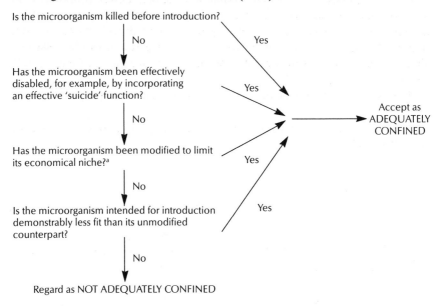

Note: [a] With respect to substrate utilisation, host range, physiological tolerance, resistance, or competitiveness.

works provide the responsible US government agencies with the foundation for a flexible, scientifically based, decision-making process. Use of the frameworks for evaluation of field tests permits the classification of an introduced organism into a risk category (NRC, 1989). The framework should also provide a valuable approach to risk assessment for regulatory agencies in other countries.

Commercial releases

In the past few years, several novel products, produced with the aid of new biotechnologies, have passed through the necessary regulatory process in several countries and have been approved for commercial sale. These have been mainly pharmaceuticals for human health care (insulin being the first such novel product released for sale). Some possible and representative environmental applications of genetically engineered organisms in agriculture are listed in Table 7.3.

Several new animal vaccines and animal growth hormones produced by

Fig. 7.9 Potential environmental effects. Appropriate questions for specific applications to be added by users of the framework for the release of genetically modified microorganisms. Source: Modified from NRC (1989).

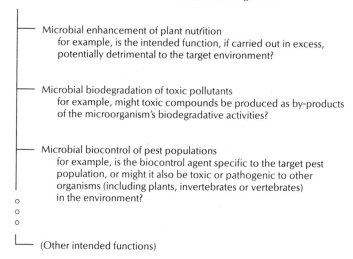

What is the intended function of the introduced microorganism?

- Microbial enhancement of plant nutrition
  for example, is the intended function, if carried out in excess, potentially detrimental to the target environment?

- Microbial biodegradation of toxic pollutants
  for example, might toxic compounds be produced as by-products of the microorganism's biodegradative activities?

- Microbial biocontrol of pest populations
  for example, is the biocontrol agent specific to the target pest population, or might it also be toxic or pathogenic to other organisms (including plants, invertebrates or vertebrates) in the environment?
  o
  o
  o

- (Other intended functions)

recombinant DNA technology are now available commercially in some countries. The recent development of a new recombinant DNA vaccine against rinderpest disease of cattle is potentially of wide applicability in Africa (Yilma *et al.*, 1988).

Several plant varieties with novel characteristics, such as insect resistance based on the presence of the *Bacillus thuringiensis* toxin gene or virus resistance based on a coat protein gene, are under field test, in at least ten countries. None are yet being sold. It is predicted that plant varieties with such novel characteristics will be one of the major commercial products for biotechnology within the next decade.

The approval for sale of 'NoGall' in Australia in February 1989 was the world's first commercial biopesticide based on a live, genetically engineered microorganism. The new biocontrol agent is a genetically modified strain of *Agrobacterium tumefaciens*. The parent strain has been used in many countries as a biocontrol agent against crown gall disease of temperate fruit trees since the mid 1970s. The modified strain has a small piece of DNA removed, so that it is no longer able to cross with the pathogenic strains of *Agrobacterium tumefaciens* in nature (Jones *et al.*, 1988; Kerr, 1989; Wright, 1989). This means that the new strain should retain the ability to control the pathogen indefinitely. The release of a novel, genetically engineered biocontrol agent in Australia may set a precedent for other similar releases elsewhere.

Table 7.3.   Some environmental applications of genetically engineered organisms in agriculture.

Microorganisms
- Bacteria as biocontrol agents against disease.
- Bacteria carrying *Bacillus thuringiensis* toxin to reduce loss of crops to black cutworm.
- Mycorrhizal fungi to increase plant growth rates by improving efficiency of root uptake of nutrients.
- Plant symbionts: nitrogen-fixing bacteria to increase nitrogen available to plants, and decrease need for fertilisers.
- Viruses as pesticides; insect viruses with narrowed host specificity or increased virulence against specific agricultural insect pests.
- Myxoma virus modified so as to restore its virulence against rabbits.
- Vaccines against animal diseases including: swine pseudorabies; swine rotavirus; vesicular stomatitis (cattle); foot-and-mouth disease (cattle).

Plants
- Herbicide resistance or tolerance.
- Resistance to plant virus diseases (e.g. potato with resistance to potato leaf roll virus).
- Pest resistance in *B. thuringiensis*-toxin-protected crops, including tobacco and tomato.
- Seeds with enhanced anti-feedant content to reduce losses to insects while in storage.
- Enhanced tolerance to environmental factors including: salt, drought, temperature, heavy metals.
- Modified protein content.

Animals
- Livestock species engineered to enhance weight gain or growth rates, reproductive performance, disease resistance, etc.
- Animals engineered to function as producers of pharmaceutical drugs, especially of mammalian compounds that require post-synthesis modifications in the cell.

Source: Modified from OTA (1988b).

Implications for IARCs

All biotechnology programs conducted at the IARCs need to abide by the biosafety regulations of the host country where the research is being conducted. Where these regulations are not yet in place, reference should be made to relevant experience elsewhere and to international guidelines such as those promulgated by the OECD, to ensure that adequate safeguards are followed. Successful testing of organisms in countries with established regulatory processes, provides firm justification for subsequent testing in other countries. Flexibility in establishing regulations will be required to accommodate rapidly evolving innovations from recombinant

Fig. 7.10 Potential relations between institutional biosafety committees of IARCs and NARSs.

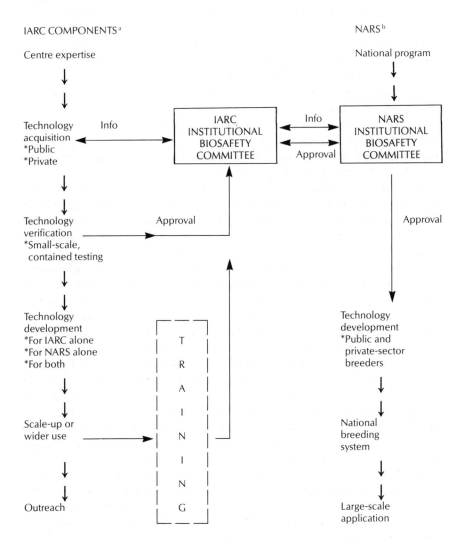

Notes: [a]IARC, International Agricultural Research Centre.
[b]NARS, National Agricultural Research System.
Source: Plucknett *et al*. (1990).

DNA research while being respectful of environmental and ethical concerns (Plucknett *et al.*, 1990).

The IARCs must deal effectively with the technical, environmental, regulatory and policy dimensions of biotechnology in order to advance the appropriate applications of the new technologies. One mechanism to ensure IARC participation in each of these areas is for each centre to establish its own Institutional Biosafety Committee (IBC), where one does not already exist. Such an IBC should work within the centre as well as with appropriate host country officials to ensure that research projects have been reviewed carefully in relation to the type of product being developed, and the environment in which it will be tested and ultimately used. While these two categories are of prime importance, the type of technology being used will continue to warrant special attention.

The IARC/IBCs would monitor, approve and regulate as necessary the development of new technologies and organisms in international research, in accordance with host country approval mechanisms and current regulatory standards recommended by the OECD. The IBC should include members from the host country. It should also include expert members knowledgeable about biosafety issues elsewhere.

It is essential that a centre be affiliated with national regulatory bodies exercising oversight over specific research activities, where such bodies exist in the host country. If not, the IARC could assist in the establishment of a functioning national review body, if requested. In the absence of a host-country national review body, appropriate international standards should be followed, with preliminary, small-scale or contained experiments conducted with industrialised country partners and in full compliance with their existing regulatory standards. The successful completion of these tests would then warrant further testing at the centre or national facility in accordance with host-country and centre approval (Plucknett *et al.*, 1990).

Possible responsibilities for an IARC-based IBC, and its proposed interactions with a national biosafety review body, are shown in Fig. 7.10. Some of these interactions will be for information while others will be for approval. A possible integration of the IBC into the path of technology development for both the IARC and the NARS is also shown in Fig. 7.10. Similarly illustrated are the various levels of co-ordination that will be required for effective collaboration regarding the acquisition and development of techniques for both contained and larger-scale testing.

The IBC must ensure that appropriate host-country approval for testing has been obtained where required, and that safety measures consistent with the technology, product and the environment have been recommended. This procedure would also help the IARCs safeguard their responsibilities to their clients, both in the host country and elsewhere.

Findings

1. National review bodies

There is a need for individual countries to establish functioning and well-informed national review bodies and institutional biosafety committees to monitor and regulate the release of genetically engineered organisms into the environment. The national body and its guidelines should be framed in order to complement and support existing regulatory agencies.

2. IARC institutional biosafety committees

There is a need for the IARCs to establish international biosafety committees, where these do not exist, and ensure that these function in accordance with host-country approval mechanisms.

3. Environmental risk assessment procedures

National review bodies and the IARCs need ready access to regular assessments of the experience and current policies of countries with functioning national review committees, particularly in relation to environmental risk assessment procedures, so that they can benefit rapidly from the experience being gained in other countries and by international organisations.

4. Project biosafety reviews

International development agencies need to ensure that biosafety reviews, using functioning national guidelines, are conducted prior to the release of genetically engineered organisms in any sponsored projects.

Chapter eight:
Patent issues

The lack of patent protection is a major disincentive for private sector investments in biotechnology in developing countries ...

Introduction

A major policy issue that will affect the application of biotechnology in agriculture is the management of intellectual property. In the USA patent protection has been granted under existing legislation to biotechnological processes (in the 1970s), and to novel microorganisms (1980), plants (1985) and animals (1988) produced using recombinant DNA technology (OTA, 1989). Several other OECD countries also offer some patent protection to biotechnological products and processes. Few non-OECD countries offer any patent protection. *The lack of patent protection is a major disincentive for private-sector investments in biotechnology, both by local private-sector companies and transnational corporations.*

Many industrialised countries and transnational companies see what they regard as inadequate protection of intellectual property rights (IPR) as a major disincentive to preparing novel products for use in the Third World. Conversely, many newly industrialised and developing countries consider that the current international patent regime works to the disadvantage of those countries with the ability to copy inventions and that these countries receive nothing in return for protecting inventions produced in industrialised countries. If no ways are found to build bridges between these divergent viewpoints, disagreements over intellectual property rights will be a major impediment to the generation of applications of biotechnology for agriculture in developing countries.

A difference in the national interests of three major country groups, (1) technology sellers, (2) technology buyers with a strong indigenous adaptive capacity, and (3) technology buyers with little research capacity, is the cause of present disagreement in the attempts to develop an international system of intellectual property rights. Strong patent laws are important to the technology sellers. For countries with no significant patent sales in other countries, the present patent conventions are not relevant. Many countries do not see the intellectual property rights requested by the industrialised countries as a 'natural right' but as a negotiating issue for which they may attain other concessions in return for opening their technology markets. Intellectual property rights are the subject of negotiation in the Uruguay round of GATT negotiations and at the World International Property Organization (WIPO) in Geneva. The issues involved in patent protection in biotechnology are described in detail in Evenson and Putman (1990) and Barton (1989).

Several countries are under considerable pressure from the USA and other OECD countries to strengthen their systems of intellectual property protection and to recognise the right of inventors from industrialised countries. In many cases, the decisions on new policies are being taken by agencies that have little input from agricultural interests.

International protection (i.e. the protection provided to an invention from one country by other countries) is now an integral part of trade negotiations, and is seen to be important to international competitiveness. For example, US firms claim that 'piracy,' i.e. the illegal infringement of intellectual property rights, is costing US firms billions of dollars in export sales. The USA is exerting considerable economic pressure on individual countries in bilateral trade negotiations for strengthened protection of intellectual property rights. South Korea and Thailand are presently reexamining their existing laws regarding intellectual property rights as a result of US pressure in trade negotiations.

The current international debate on 'patenting life' stems both from: (1) the economic debate as to what is an equitable international system for recognising the contribution of an invention, while not stifling the development of inventive capacity elsewhere, particularly in the Third World; and (2) the ethical and legal debates as to whether living organisms produced by new biotechnologies are patentable subject matter or part of the common heritage of mankind. Many countries specifically exclude agricultural inventions from patent protection on the grounds that they provide 'basic needs' for food and health care for the population.

Types of inventions

The term 'invention' broadly covers any new process, device, chemical composition or thing which has been developed, devised or discovered. Inventions stemming from agricultural research can be grouped as follows: (1) mechanical/electrical; (2) chemical; (3) biological/genetic; and (4) other (e.g. computer programs). Evenson and Putman (1990) have described these categories as follows:

Mechanical/electrical inventions primarily include new machinery for planting, cultivating and harvesting agricultural crops, as well as new machinery and equipment used in animal husbandry. These inventions

Table 8.1.   Legal systems for protecting inventions.

Seed and breed certification systems

These normally require that seed and animals be marketed with sufficient labelling to identify the origin of the seed or animal and give its genetic heritage. Such certification operates like a trademark to prevent others from trading on the reputation that a breeder establishes with a new plant or animal variety. These systems do not usually prevent others from using and selling the same plant and animal varieties as long as they do not misrepresent it.

Plant patent and variety protection

Plant patent and variety protection systems provide plant breeders with limited rights to exclude others from commercialising new plant varieties they have developed. To be protected a plant variety must be uniform, stable and distinct from all known varieties. Typically, some type of control depository is provided to preserve seed samples for the plant varieties being protected, in addition to the requirements for verbal and pictorial description. Most plant protection laws exempt experimental breeding from infringement claims in order to encourage the use of protected varieties in further breeding/research programs. Also, other provisions may be included, such as in the system in the USA which permits farmers to save seed from their crop for their own use on their farm and sell it to other farmers, for purposes other than seed reproduction.

Invention patents

An invention patent system, which governs the usual type of patent, gives the inventor the right to exclude others from practising the invention for a certain time period (usually 15–20 years). Invention patent systems traditionally require that an application for a patent must include an 'enabling disclosure' which sufficiently describes the invention so that others skilled in the same technical field can reproduce it successfully. Patent laws thus encourage early publication of the invention in return for granting a limited monopoly. From the perspective of economic efficiency, such disclosure is preferable to trade secrecy.

An invention patent is not an exclusive right to practise a particular invention, but rather a right to exclude others from practising the invention, within the scope of the exclusion defined by the claims, which describe the novel contributions made by the inventor.

To be valid, an invention patent must disclose an invention that is *novel, useful, and unobvious in view of the prior art*. An invention must be novel in the sense that it has not previously been published, exhibited or otherwise described (except within the period immediately preceding the application, and then only by the applicant). As to its utility, the invention must be capable of industrial or agricultural application, and not be purely ornamental.

Petty patents (utility models)

Petty patents (sometimes called utility models) are similar to invention patents in that they give the inventor the right to exclude others from practising the invention for some period of time. They differ from the development patents in requiring novelty and utility, without requiring any inventive step above the prior art. Thus, petty patents preserve rights to minor variations of known devices rather than to major technical innovations having broad adaptability. In those countries that provide them, petty patent protection is usually granted for a much more limited time than is the case for invention patents, and usually only to their own citizens. Also, since the

Table 8.1. cont.

---

existence of an inventive step is not a requirement, such systems cost less to administer than do most invention patent systems.

Petty patent systems tend to stimulate domestic developments. Minor adaptations of machinery and other developments to accommodate local conditions help the local economy but may not be valuable abroad. Nationals of developing countries with petty patent systems are more likely, therefore, to utilise petty patents rather than the more costly and difficult-to-obtain invention patents.

Inventor's certificates

An inventor's certificate is a notice given in socialist countries that entitles an inventor to receive compensation for his or her invention, which as a matter of property belongs to the state. This nonmarket alternative to the standard patent aims to reward the inventor while removing his or her monopolistic control over the invention.

Industrial design patents

Industrial design patents provide protection to designs as opposed to inventions per se. They provide narrower protection, somewhat similar to that provided by copyrights.

Copyrights

Copyright prevents unlicensed copying of works of art or an author's writings. The copyright is quite literally limited to copying the publication and does not preclude the use of the information contained therein.

Trade secrecy

For inventions which can be maintained in secrecy, such as manufacturing processes that are not readily apparent in the marketed products, trade secrecy contracts can prohibit anyone (primarily ex-employees and collaborators) from disclosing secrets of manufacture and the like to competitors. However, they provide little protection against unauthorised disclosure by persons not party to the contract. Moreover, no protection is available once secrecy is lost.

Source: Evenson and Putman (1990).

frequently require small modifications in order to be transferred from one region to another.

Chemical inventions include fertilisers, insecticides and pharmaceuticals for animals. These inventions are usually more widely adaptable than are mechanical/electrical inventions.

Biological/genetic inventions include improved plant varieties and animal breeds, the products of new genetic engineering processes including recombinant DNA technology, and the new processes themselves. The products of biological/genetic research, especially plants, are usually quite limited in their geographic adaptability due to varying soil and climatic conditions and different pest and disease profiles.

'Other' inventions for agriculture include computer programs and general

business systems for improving the conduct of agricultural research or farm management.

Legal systems for protecting inventions

Several different legal systems are available for securing private rights to new developments. The main legal systems relevant to the protection of new developments (and intellectual property) in agriculture are: (1) seed and breed certificates; (2) plant patent and variety protection; (3) invention patents; (4) petty patents (utility models); (5) inventor's certificates; (6) industrial design patents; (7) copyrights; and (8) trade secrecy.

All these systems provide some type of legally enforceable right to restrict the use of the development by someone other than the creator and his/her licencees/assignees. The key characteristics of each of these systems are given in Table 8.1.

Availability of patent or other protection

In order to summarise the relative scope of protectable subject matter in various countries, consider the R and D program of a hypothetical Amalgamated Agricultural Research Corporation (AARC). For the purpose of this discussion, it is assumed that all the research results were obtained exclusively by AARC, using AARC funds, never before existing or described anywhere in the world, and were better than the state-of-the-art (Evenson and Putman, 1990).

Several developments from the AARC research and development program are now being considered by the AARC Board for potential commercialisation. They came from the categories of:

1. *Mechanical/electrical* (new machinery).
2. *Chemical* (a new fertiliser, insecticide, herbicide and an animal pharmaceutical).
3. *Biological/genetic* (new plant varieties, a new animal breed, a new bacterial strain produced by recombinant DNA techniques and a recombinant DNA viral vaccine).

Further details of the developments are given in Table 8.2. The current availability of protection for the intellectual property rights of each of these hypothetical developments in selected countries is listed in Table 8.3.

The degree of protection given to intellectual property rights of agriculture-related developments varies considerably among countries. Of particular interest are those entires in Table 8.3 that denote that the development in question is specifically prohibited from being patented or

Table 8.2. Fourteen different developments from a hypothetical Amalgamated Agricultural Research Corporation (AARC).

Invention	Description
Mechanical/electrical developments	
1. AARC plough I:	Basic innovation in ploughs which is adaptable to a wide range of soil and climatic conditions.
2. AARC plough II:	Minor modification for adapting a plough to be utilised in a specific localised soil condition.
Chemical developments	
3. AARC chemical fertiliser:	New fertiliser chemical compound which optimises tolerance to extremes in soil moisture conditions.
4. AARC insecticide:	New and useful chemical compound insecticide.
5. AARC herbicide:	New and useful chemical compound herbicide.
6. AARC pharmaceutical for animals:	Chemical compound pharmaceutical for controlling disease in farm animals.
Biological/genetic developments	
7. AARC soybean:	New improved soybean variety developed in a plant breeding program.
8. AARC maize:	New hybrid maize seed variety developed in a plant breeding program, with AARC retaining control over the hybrid parents.
9. AARC rose:	New variety of asexually reproducible rose.
10. AARC beef cattle:	New pure breed of beef cattle developed in a selective breeding program.
11. AARC bacterium:	New and improved nitrogen-fixing strain of bacteria developed using recombinant DNA techniques.
12. AARC live virus vaccine:	New strain of virus developed using recombinant DNA techniques to be used as a vaccine for animals.
Other developments	
13. AARC computer program:	A new and improved computer program that determines the optimal mix of chemicals in the AARC Chemical Fertiliser (development no.3).
14. AARC accounting system:	A new and improved accounting system for allocating resources, adaptable to any large agricultural research organisation.

Source: Evenson and Putman (1990).

Table 8.3. Availability of patent or variety protection for different categories of developments in selected countries.

Country	Mechanical/electrical		Chemical				Biological/Genetic						Other	
	Plough I major	Plough II minor	Ferti-liser	Insecti-cide	Herbi-cide	Chemical vaccine	Soybean	Maize hybrid	Rose	Beef cattle	Nitrogen-fixing bacteria[a]	Live virus vaccine[a]	Computer program	Accounting systems
Australia	yes	no	yes	yes	yes	no	yes	yes	yes	no	yes	no	no[b]	no[b]
North America														
Canada	yes	yes	yes	yes	yes	yes	no	no	no	no	yes	yes?	no[b]	no[b]
USA	yes	yes	yes	yes	yes	yes	yes	yes?	yes	no	yes	yes	yes?	no
European Community														
France	yes	no	yes	yes	yes	yes	yes	yes	yes	no	yes	yes?	no[b]	no[b]
Netherlands	yes	no	yes	yes	yes	yes	yes	yes	yes	no	yes	yes	no[b]	no[b]
UK	yes	no	yes	yes	yes	yes	yes	yes	yes	no	yes	yes?	no[b]	no[b]
West Germany	yes	no	yes	yes	yes	yes	yes	no?	yes	?	yes	yes	no	no
Asia														
Bangladesh 1,2	no	no	no	no	no	no	no[c]	no[c]	no[c]	no[c]	?	?	no[c]	no[c]
China	yes	no?	no?	no?	no?	no?	no	no	no	no	?	no	no	no
India	no[c]	no	no[d]	no[d]	no[d]	no[d]	no	no	no	no[c]	no[c]	no[c]	no	no
Indonesia 2	—	—	—	—	—	—	—	—	—	—	—	no[c]	—	—
Korea (South)	yes	yes	no[d]	no[d]	no[d]	no[c]	no[c]	no[c]	yes	no[c]	yes	no[d]	no[c]	no
Malaysia 1,2	yes	no	yes	yes	yes	yes	no[c]	no[c]	(see note)	no[c]	no	no	no	no[c]
Nepal 4	yes	no	yes	yes	yes	yes							no	no
Pakistan 3	yes	no	yes	yes	yes	yes	no	no[c]	no[c]	no[c]	yes	yes	no	no[d]
Philippines 5	yes	yes	yes?	yes?	yes?	no	no	no	no	no	yes	yes	yes	no[d]
Singapore 1,3	yes	no	yes	yes	yes	yes	no[c]	no[c]	no[c]	no[c]	no	no	no[d]	no
Sri Lanka 3	yes	no	no[d]	yes	yes	yes	no[c]	no[c]	no[c]	no[c]	no	no	no[c]	no[d]
Taiwan	yes	yes	no[d]	no[d]	no[d]	no[d]	no[c]	no[c]	no	no[c]	no	no[c]	no	no[d]
Thailand	no[c]	no	yes	yes	yes	no[d]	no[c]	no[c]	no[c]	no[c]	no[c]	no[c]	no[c]	no
Latin America														
Argentina	yes	no	yes	yes	yes	no[c]	yes	yes	yes	no	yes	no[c]	no[c]	no[c]
Bolivia	yes	no	no[d]	no[d]	no[d]	no[d]	no	no	no	no	no	no[c]	no[c]	no[c]
Brazil	yes	yes	no[d]	no[d]	no[d]	no[c]	no[c]	no[c]	no	no[c]	no	no[c]	no[c]	no[c]
Chile	yes	no	no[d]	no[c]	no[c]	no[d]	no[c]	no[c]	no[c]	no[c]	no	no[c]	no[c]	no[c]
Colombia	yes	no	no[d]	no[d]	no[d]	no[d]	no[c]	no[c]	no[c]	no[c]	no	no[c]	no	no
Costa Rica	yes	no	yes	yes	yes	no[c]	no	no	no	no	no	no[c]	no	no

Dominican Republic	yes	no	no	yes	yes	yes	no	no	no	no	no	no	no[c]	no[c]
Ecuador 7	yes	no	yes	yes	yes	no[c]	no[c]	no[c]	no[c]	no	no[c]	yes	no[c]	yes
Guyana	yes	no	yes	yes	yes	yes	no	no	no	no	no	no	no	no
Haiti 3	yes	no	yes	yes	yes	yes	no	no	no	no	no	no	no	no
Jamaica	yes	no	yes	yes	yes	yes	no	no	no	no	no	no	no	no
Mexico	yes	no	no[c]	no[c]	no[c]	no[c]	no[c]	no[c]	no[c]	no[c]	no[c]	no[c]	no[c]	no[c]
Nicaragua	yes	no	no[d]	no[d]	no[d]	no[d]	no	no	no	no	no	no	no[d]	no
Panama 5	yes	no	yes	yes	yes	yes	no	no	no	no	no[c]	no	no	no
Paraguay	yes	no	no[d]	no[d]	no[d]	no[d]	no[c]	no[c]	no[c]	no[c]	no[c]	no[c]	no[c]	no[c]
Peru	yes	no	no[d]	no[d]	no[d]	no[d]	no[c]	no[c]	no[c]	no[c]	no	no	no	no
Surinam	yes	no	no[d]	no[d]	no[d]	no[d]	no[c]	no[c]	no[c]	no	no	no	no[c]	no[c]
Uruguay	yes	yes	no[d]	no[d]	no[d]	no[d]	no[c]	no[c]	no[c]	no	no[c]	yes	yes	no[c]
Venezuela	yes	no	no[d]	no[d]	no[d]	no[d]	no	no	no	no	no[c]	no	no	no
Near and Middle East														
Arabia 1	yes	no	yes	yes	yes	yes	no[c]	no[c]	no[c]	no	no	no[c]	no[c]	no[c]
Bahrain 1	yes	no	yes	yes	yes	yes	no[c]	no[c]	no[c]	no[c]	no	no[c]	no[c]	no[c]
Egypt	yes	no	yes	yes	yes	no[d]	no	no	no	no	no	no	no	no
Iran	yes	no	yes	yes	yes	no[d]	no	no	no	no	yes	yes	yes	yes
Iraq	yes	no	yes	yes	yes	no[d]	no	no	no	no	yes	no	no	no
Israel	yes	no	yes	yes	yes	yes	no	no	no	yes	yes	yes	yes	yes
Jordan	yes	no	yes	yes	yes	yes	no	no	no	yes	no[c]	yes	no[c]	yes
Kuwait	yes	no	yes	yes	yes	no[d]	no	no	no	no	no[c]	no	yes	yes
Lebanon	yes	no	yes	yes	yes	no[d]	no	no	no	no	no	yes	no	no
Syria	yes	no	yes	yes	yes	no[d]	no	no	no	yes	no	yes	yes	yes
Africa														
African Intellectural Property Organisation 3,6	yes	no	yes	yes	yes	yes	no[c]	no[c]	no[c]	no	no[c]	yes	no	no[c]
Algeria 3,7	yes	no	yes	yes	yes	yes	no[c]	no[c]	no[c]	no	no	yes	no	no[c]
Botswana, Lesotho 3	yes	no	no	no	no	no	no[c]	no[c]	no[c]	no	no	no	no	no[c]
Burundi, Rwanda, Zaire 5	yes	no	yes	yes	yes	yes	no	no	no	no	no	yes	no	no
Ghana 3	yes	no	yes	yes	yes	yes	no	no	no	no	no[c]	no[c]	no[c]	no[c]
Kenya 3	yes	no	yes	yes	yes	no[d]	no[c]	no[c]	no[c]	no	no[c]	no[c]	no	no[c]
Liberia	yes	no	yes	yes	yes	yes	no[c]	no[c]	no[c]	no	no	no[c]	no	no
Libya	yes	no	yes	yes	yes	no[d]	no	no	no	no	no[c]	no[c]	no[c]	no[c]
Malawi, Zambia, Zimbabwe	yes	no	yes	yes	yes	yes	no	no	yes	yes	no[c]	yes	yes	no[c]
Nigeria 3	yes	no	yes	yes	yes	yes	no[c]	no[c]	no[c]	no	no	no	no	no
Sierra Leone 3	yes	no	yes	yes	yes	yes	no[c]	no[c]	no[c]	no	no	no[c]	no[c]	no[c]

Table 8.3. cont.

Country	Mechanical/electrical		Chemical				Biological/Genetic						Other	
	Plough I major	Plough II minor	Ferti- liser	Insecti- cide	Herbi- cide	Chemical vaccine	Soybean	Maize hybrid	Rose	Beef cattle	Nitrogen- fixing bacteria[a]	Live virus vaccine[a]	Computer program[a]	Accounting systems
SW Africa 5	yes	no	yes	yes	yes	yes	no	no	no	no	no	no	no	no
Sudan	yes	no	yes	yes	yes	yes	no	no	no	no	no	no	no	no
Tanzania	yes	no	yes	yes	yes	yes	no	no	no	no	no	no	no	no
Tunisia	yes	no	yes	yes	yes	no[d]	no[c]	no[d]	no	no	no	no	no	no

Notes: This table summarises the data that record principal constituents of each country's patent laws.

[a]Biotechnology-based product, developed using recombinant DNA technology.

[b]Patent protection not available but copyright as other similar protection available.

[c]This invention is specifically excluded from patent protection by national law.

[d]Although this chemical substance is specifically excluded from patent protection by national law, the process used to produce the substance is not excluded.

1. British patent law is assumed to hold in this country, owing to the provisions in its laws. British patent applications (whether or not by British citizens) have priority. In practice, a prior British patent is routinely granted approval in this country at the applicant's request. See Chapter 37 of the Patents Act of 1977 of Great Britain. The UK prohibits the patenting of microbial processes or products for use on humans or animals. Ghana independently prohibits patents on pharmaceutical and medical substances.

2. This country has no patent act of its own.

3. 'Microbiological processes and the products of such processes' are patentable. Whether this protection extends to microorganisms per se is not known and will depend on the interpretations of the domestic courts. In the absence of specific indications to the contrary, we have assumed that the nitrogen-fixing bacteria and the live virus vaccine are not patentable under these circumstances.

4. A patent is granted to a foreign inventor if he/she has obtained a patent in his/her own country and any three other countries. Presumably, patentability standards in those countries apply.

5. Other than meeting public standards of health and morality, no other criteria for patentability are cited. In general, we take mechanical, chemical and electrical inventions to be patentable, and others to be unpatentable. In the Philippines, US law is assumed.

6. The following countries are signatories to the Libreville Agreement of 1962, which establishes the African Intellectual Property Organisation: Benin, Cameroon, Central African Republic, Chad, Congo, Gabon, Côte d'Ivoire, Malagasy Republic, Mauritania, Niger, Senegal, Togo and Burkina Faso. The revised agreement of 1977 has been signed by Cameroon, the CAF, Gabon, Côte d'Ivoire, Mauritania, Niger, Senegal, and Togo. In the absence of laws to the contrary, we apply the revised standards to the other countries as well.

7. The Inventor is entitled to state indemnification for the rights to some or all of these inventions. In this case, he/she does not own the rights himself/herself.

8. Food and chemical patents require mandatory licensing.

Source: Evenson and Putman (1990).

otherwise protected. Agriculture is unusual among traditional production industries in that many agricultural developments are either ineligible for protection or have an indeterminate status under most national patent acts (Evenson and Putman, 1990).

As to biological/genetic developments, most countries specifically exclude plant varieties and animal species from protection, either as such or excluding foodstuffs. Among newly industrialised countries, only Argentina and South Korea make provision for plants of any kind to be patented, and only Argentina permits sexually reproduced plants to be patented.

Recent interpretation of the US patent system (in 1987) allows the patenting of plant varieties. Although there are legal systems in some countries (e.g. Hungary and Romania) which provide breeder's rights to new animal breeds, none of the countries listed in Table 8.3 has such a provision, so none would protect AARC's new strain of beef cattle. The recent decision taken in the USA (in 1988) to issue a patent on a genetically engineered mouse for use as an experimental tool in cancer research has set a precedent for the patenting of novel animals produced using recombinant DNA technology (OTA, 1989). The European Commission is now considering the extension of its patent protection in biotechnology to plants and animals (EC, 1988).

With regard to the patenting of genetically engineered microorganisms, the USA, Argentina, South Korea, Israel, Jordan, the Philippines, Syria and Zimbabwe, allow the patenting of the genetically engineered bacteria, as a novel microorganism not occurring in nature. With regard to the live virus vaccine produced using recombinant DNA technology, the USA, Pakistan, the Philippines, Israel and Jordan would allow it to be patented. Argentina and South Korea specifically disallow the live virus vaccine on the grounds that it is a medicine.

The 'other development' category, namely the computer program and the accounting system, do not generally receive patent protection or the like under most legal systems. The following countries give patent protection to computer programs: Ecuador, Iran, Japan, Jordan, Kuwait, Pakistan, Syria and Uruguay.

Role of international conventions on intellectual property

In keeping with the treatment of developments as 'intellectual property,' most countries are party to at least one international agreement, the intent of which is to protect an inventor's rights to his/her invention from country to country. The major international conventions relevant to agriculture are the Paris Union Convention; the Patent Co-operation Treaty; the European Convention; the Budapest Convention; and the International Union for the Protection of New Varieties of Plants (the UPOV Convention). The current membership of the conventions is listed in Table 8.4.

Table 8.4.   Membership of international conventions on intellectual property.

Convention	No.	Member countries
Paris Union Convention[a]	97	
European Patent Convention	13	Austria, Belgium, France, Federal Republic of Germany, Greece, Italy, Liechtenstein, Luxembourg, Netherlands, Spain, Sweden, Switzerland, United Kingdom
Patent Co-operation Treaty	40	Australia, Austria, Barbados, Belgium, Benin, Brazil, Bulgaria, Cameroon, Central African Republic, Chad, Congo, Democratic People's Republic of Korea, Denmark, Finland, France, Gabon, Federal Republic of Germany, Hungary, Italy, Japan, Liechtenstein, Luxembourg, Madagascar, Malawi, Mali, Mauritania, Monaco, Netherlands, Norway, Republic of Korea, Romania, Senegal, Soviet Union, Sri Lanka, Sudan, Sweden, Switzerland, Togo, United Kingdom, United States of America
Budapest Treaty	22	Australia, Austria, Belgium, Bulgaria, Denmark, France, Finland, Federal Republic of Germany, Hungary, Italy, Japan, Liechtenstein, Netherlands, Norway, Philippines, Soviet Union, South Korea, Spain, Sweden, Switzerland, United Kingdom, United States of America
UPOV Convention	17	Australia, Belgium, Denmark, France, Federal Republic of Germany, Hungary, Ireland, Israel, Italy, Japan, Netherlands, New Zealand, South Africa, Spain, Sweden, Switzerland, United Kingdom, United States of America

Note: [a]Approximately 100 countries belong to the Paris Union Convention.
Sources: Beier *et al.* (1985); OTA (1989).

Paris Union Convention

The Paris Convention stipulates that member countries will grant patents to foreign nationals on the same terms as it grants patents to its own citizens. It defines patents broadly as including 'the various kinds of industrial patents recognized by the laws of countries of the Union, such as patents of importation, patents and certificates of addition, etc'. This broad language permits any of the forms of industrial patents granted under the law of member countries to be included. The Paris convention was established in 1883 (in Paris), and modified in 1925 (The Hague), 1934 (London), 1958 (Lisbon) and 1967 (Stockholm). It is the intellectual property agreement most widely subscribed to, having nearly 100 member countries. India and Thailand are notably absent.

The Convention permits member nations to enter into separate agreements for the protection of intellectual property as long as they do not contravene the provisions of the convention. Under this provision several other multinational agreements (e.g. the European Patent Convention and the Budapest Treaty) have been concluded.

Patent Co-operation Treaty

This worldwide convention is open to any member of the Paris Union. It was established in 1978, and by 1988 applied to 40 member countries. The Patent Co-operation Treaty deals only with procedural requirements related to co-ordinating and simplifying the filing, searching and publication of international patent applications. Its primary accomplishment is to obtain an international patent search at an early stage, and thus an early indication of patentability. About half of the signatories are newly industrialised or developing countries.

European Patent Convention

The European Patent Convention establishes a system of law common to all 13 member countries to provide for a uniform procedural system for the centralised filing, searching and examination for European patents. It took effect in 1977, following its formulation in 1973. Each granted patent takes effect only in those member countries which it designates.

Budapest Treaty

The Budapest Treaty on the International Recognition of the Deposit of Microorganisms for the Purposes of Patent Procedure deals with the logistics of patenting relevant to microbes and microbial processes. It provides that member states recognise for their own patent procedures deposits of microbes made in the appropriate form in certified depositories in other member countries. Twenty-two member countries rely upon 18 depository institutions in member countries (as of December 1987). The treaty was entered into in 1980.

UPOV Convention

In 1957 an international conference was held in Paris for the purpose of drafting a convention for protecting new plant varieties. It was signed in

1961, entered into force in 1968, and was last modified in 1978. The objective of the convention is 'to recognize and to ensure to the breeder of a new plant variety ... the right to a special title of protection or of a patent'. Seventeen member countries provide reciprocal protection. Each protected variety must have a specific, unique name for marketing purposes. The plants may be sexually as well as asexually reproduced (which gives protection to hybrid varieties). No developing countries belong to this treaty.

International debate on patenting life

Patent availability

Private-sector companies have been investing substantially in R and D for modern biotechnology, largely because they perceive that there are opportunities for the proprietary protection of the new products and processes. Protection of intellectual property rights is demonstrably one of the most effective ways of stimulating private-sector innovation in the development of biotechnology applications. This is shown by the rapidly increasing number of patents issued in biotechnology, which in the USA have almost doubled between 1983 and 1987 (OTA, 1989).

The central issue in the debate on 'patenting life' is that patents have been obtainable traditionally only for manufactured inventions, not for naturally occurring subject matter, which belongs in the public domain. The argument in favour of extending patent protection in biotechnology is that the processes, substances and organisms that fall under the rubric of biotechnology do not occur naturally, or, if the substance also occurs naturally, the process used to produce it in commercial quantities is made by people (Evenson and Putman, 1990). The arguments against are based on the concept of the naturally occurring germplasm being part of the common heritage of mankind. These arguments are well summarised by Fowler *et al.* (1988), and Farrington (1989).

In recent years, the industrialised countries (particularly the USA and Japan) have interpreted their existing laws in such a manner as to give greater protection to intellectual property rights in biotechnology. The precedent for this was set in the Diamond *v.* Chakrabarty patent law case in the US Supreme Court in 1980, where the court overruled the US Patent Officer by finding that microorganisms are patentable subject matter within the patent law then in force. The court ruled that the microbe under consideration, produced using recombinant DNA technology, met the three stipulations of US Patent Law, namely that it was a composition of matter that was useful, nonobvious and novel.

This decision was upheld and extended in subsequent rulings in the USA in 1985 to apply to plants produced using recombinant DNA techniques. In 1988 the US Patent Office issued the first-ever animal patent to Harvard University for a transgenic mouse designed to be used in cancer research. At the time the patent was issued in 1988, there were approximately 20 additional applications pending for animal patents in the USA (OTA, 1989).

In early 1989 information was publicly available on seven animal patent applications before the European Patent Office (EPO) in The Hague. The European Patent Convention stipulates that patents will not be issued for plant or animal varieties and essentially biological processes for the production of plants and animals (with the exception of microbial processes or the products thereof). This means that the EPO approach to patenting living organisms will have to be less direct, and possibly more dependent on technical details than the approach taken in the USA (EPO, 1988).

The Council of the European Community is presently considering a proposal for a Council directive on the legal protection of biotechnological inventions (EC, 1988). The aim is to unify the patent systems for biotechnological products and processes within the member countries of the European Community. The draft convention states that an invention shall not be held unpatentable for the sole reason that it is composed of living matter. The EC is considering the introduction of common patent protection in biotechnology in order to be more competitive with the USA and Japan, which currently offer stronger patent protection in this area (EC, 1988). The status of patent protection in biotechnology in all OECD countries was reviewed by Beier *et al.* (1985), and more recently by Byrne (1989).

An assessment of the current status of patent protection in biotechnology has been published by the US Congress Office of Technology Assessment (OTA, 1989). It focuses mainly on recent developments in the USA, particularly with regard to the patenting of novel animals. It also includes an analysis of the situation in several countries in Western Europe (Belgium, the Federal Republic of Germany, France, Switzerland, the United Kingdom); Eastern Europe (Bulgaria, Czechoslovakia, the German Democratic Republic, Hungary, Poland, Romania, Yugoslavia); Asia (China, Japan), North and South America (Argentina, Brazil, Canada, Chile); and Australia.

Although many biotechnology processes have been patented, few of those patents have yet been challenged in the courts. There is some concern among private companies that some of the process patents especially may not stand up if challenged.

Technology sellers *v.* technology buyers

The major policy issue revolves around the divergence in attitude towards intellectual property rights taken by buyers *v.* sellers of inventions.

The perspective of countries on the desirability or otherwise of international protection of intellectual property rights depends on whether a country is: (1) *a technology seller* (e.g. USA, Japan); (2) *a technology buyer and an adapter of imported technology,* with large internal markets (e.g. Brazil); or (3) *a technology buyer, with no local adaptive capacity* and small internal markets (e.g. Ghana). The technology sellers favour international protection, while the technology adapters and the importers do not see international protection as a 'right,' but a point of negotiation, on which they would allow the technology sellers access to their internal markets.

In biotechnology, the technology sellers are mainly the OECD countries, particularly Japan and the USA. Thus the international debate on patent protection in biotechnology is primarily a debate between the OECD countries on one hand and the developing countries on the other. This debate is occurring within the Uruguay round of GATT multilateral trade negotiations; in WIPO and in bilateral trade negotiations, particularly with the USA. It is being conducted mainly without input from agricultural interests, and with limited input from many developing countries.

A country can be seen as having the following general interests:

1. Stimulating national invention at the lowest cost.
2. Purchasing technology from abroad at the lowest possible cost.
3. Protecting the interests of its sellers of technology abroad.

Strong local patent laws and faithful adherence to the 'Paris Convention' works well for countries that are significant technology sellers. (The Paris Convention stipulates that member countries will grant patents to foreign nationals on the same terms as it grants patents to its own citizens.) This position is advanced by the USA (especially in biotechnology), where it is one of relatively few sellers of technology. Adherence to this position is also advisable for countries expecting to become technology sellers soon (Evenson and Putman, 1990).

For countries with a local adaptive inventing or copying capacity, but no significant sales abroad, the Paris Convention does not work. These countries are buyers of technology. They want foreign firms to provide inventions at low cost and they want these inventions to serve as the basis for their own adaptive inventions. They do not see the intellectual property rights of foreign countries as a natural right, but rather as a negotiated right whereby technology buyers get something in return for opening their technology markets to the technology sellers. The Paris Convention currently gives them little in return for protecting the rights of foreign firms. Hence, there is a widespread tendency to 'pirate' (i.e. to maintain the system but subtly to violate it).

For countries with little local private-sector inventive capacity, there is reason to stimulate the development of such capacity. Most of their technology purchase is in the form of interlinked contracts (inventing rights are purchased in combination with technology consulting). Strong intellectual property rights probably lower the cost of negotiating these interlinked contracts, hence they lower the cost of technology transfer to such countries (Evenson and Putman, 1990).

Biotechnology in industrialised countries is being developed largely by private-sector investments. The predominant reason for this new, high level of private investment is the fact that many of the novel products and processes are able to be patented, especially in the USA, Japan and several European countries. Thus private firms perceive that they are able to realise financial returns on their research investments in biotechnology.

The availability of patent protection for biological/genetic products is much less common outside the USA and other OECD countries. Indeed, several countries (notably Brazil and India) specifically exclude biological products from patent protection. The only countries outside the OECD member countries which currently offer some patent protection to novel bilogical/genetic products are Pakistan, the Philippines and South Korea, Argentina, Zimbabwe, and Israel, Jordan and Syria (Table 8.3).

The lack of patent protection in most developing countries is a major disincentive for private-sector investments in biotechnology, both by local private-sector companies, and by transnational companies. This may have future detrimental effects on private-sector investments in agriculture.

The role of the intellectual property debate, relative to other trade-related issues, is being considered in the Uruguay round of GATT negotiations. The mid-term agreements reached in April 1989 on the Uruguay round of GATT negotiations have broken, at least temporarily, the debate between the developing and industrialised countries as to whether the responsibility for trade-related rules for intellectual property should be with GATT or with WIPO. The USA favours having the responsibility with GATT, since it considers it has some power to enforce agreed rules. The developing countries generally favour giving the responsibility to WIPO. Work will now begin within the GATT negotiations on organising new rules for protecting intellectual property, leaving the discussion over who should apply them until later.

The advantages of the provision of intellectual property rights in biotechnology in developing countries would be to encourage: (1) the development of local research capacity; (2) greater local private-sector investments in biotechnology; (3) in-country investments by transnational companies to develop specific products for local markets; and (4) the possibility of using intellectual property rights in negotiations for improved access to overseas markets for export commodities. The major disadvantage is that it involves giving proprietary protection to living organisms which some consider to be part of the common heritage of mankind. Each country needs to weigh the benefits and costs of intellectual property rights in biotechnology, and frame its policies accordingly.

A farmer in a developing country may obtain a new plant variety by several different routes: (1) by public-sector development; (2) by a local firm's development, if that firm has the incentive to develop the technology – this incentive may depend on patent issues; (3) by a local firm's licence from a firm based in an industrialised country, if the foreign firm is confident enough of the local intellectual property and technology transfer systems; (4) by importing hybrids (but not the parental lines) on a season-by-season basis if the foreign firm is not confident enough of the local intellectual property system to import the parental lines (Barton, 1989).

Issues for IARCs and NARSs

There is a need to facilitate greater access by individual countries to biotechnological products and processes, while protecting the legitimate interests of their inventors. This could be done by licensing arrangements and other contractual agreements. International development agencies and the IARCs could facilitate such access, by negotiating (possibly on behalf of a consortium of countries) for the licensing of new technologies to apply to commodities important in the Third world.

It is possible that private companies may be able to gain sufficient protection for their proprietary technologies via suitable contractual arrangements that they will be willing to undertake co-operative development efforts with entities in developing countries.

Several countries are making changes to their policies on intellectual property rights that will affect IARC and NARS roles and concerns. It will be difficult for the IARCs to continue to simply maintain a watching brief on intellectual property rights, as is mainly the case at present.

Evenson and Putman (1990) identify several issues of concern to NARSs and the IARCs. These are: (1) private-firm franchises for research; (2) IPR protection of the germplasm component of invention; (3) public-sector provision of germplasm to private companies; (4) private-firm germplasm secrecy; and (5) new laws.

These issues span both the technology-adapting and the technology-importing countries. They become more important as the private inventive sector grows in importance in a particular country. They are not of great moment in the many countries that are at present technology importers, except that laws passed now may be the wrong laws for future development of local inventive capacity.

Private firm franchises for research – Private firms would like monopoly rights to supply technology where possible. In general, this is not a good idea. No public system should step aside and allow a complete monopoly, except in certain small sectors. Public systems have a responsibility to provide public good and to encourage competition.

Germplasm protection – Some advocates of strong intellectual property rights would extend protection of a plant or animal invention to the progeny of the associated plant or animal. This is a large extension of intellectual property rights beyond plant variety protection which currently does not limit the use of a protected variety for research (and parentage). The conceptual germplasm content of a patented invention generally is not protected by patent law. The concept embedded in an invention can be applied to another invention without infringing the original patent. IARCs and NARSs have a strong interest in seeing that germplasm protection is not strengthened to the point where research is inhibited.

Provision of public-sector germplasm to private companies – IARCs have a comparative advantage in the supply of germplasm to NARSs. Should they also supply germplasm to private firms? Yes, unless the firm has monopoly power. For example, CIMMYT (Centro Internacional de Mejoramiento de Maiz y Trigo) and IITA (International Institute of Tropical Agriculture) could make maize genetic materials available to all private firms and all NARSs, and encourage competition.

Private company trade secrecy – If private companies are to invest in R and D, they either use trade secrecy or intellectual property rights as incentive systems. Secrecy is difficult for research programs, yet part of private firms' inventive strategy is to hold elite lines, etc., in secrecy. This is not a problem if the IARCs and NARSs maintain nonsecret collections and encourage competition among the firms.

New laws – Petty patent (utility model) laws work well to stimulate local adaptive inventions. They can and should be strengthened in developing countries, to stimulate the development of the domestic R and D infrastructure. Such development has large payoffs in agriculture, where minor adaptations of known technology are necessary for successful transfer to local situations.

Patenting of IARC inventions

The IARCs could patent their significant discoveries in biotechnology, and then license (freely, if appropriate) these for use by NARSs and other collaborators. This action could be defensive patenting, to protect the IARCs against third parties patenting their inventions. It could also be offensive patenting, in terms of earning royalties for IARC inventions and using these to fund further research and development.

Patenting of IARC inventions would also facilitate the negotiation of substantive collaborative activities with private-sector companies. International development agencies and IARCs could also negotiate access to new technologies on behalf of consortia of countries, particularly the small technology importers with limited capacity to adapt technology themselves. This situation exists particularly in Africa, where there are many small countries with little local adaptive capacity at present, that would benefit from greater access to new technologies.

Findings

1. Model patent systems

Individual countries need to consider the benefits and costs of intellectual property rights in relation to biotechnology, and, where the benefits outweigh the costs, design patent systems tailored to their requirements.

2. Trade negotiations

Individual countries need to consider their position on negotiating intellectual property rights in exchange for improved trade opportunities.

3. Patent policy information

Information and advice should be made available to staff of NARSs, IARCs and international development agencies on the issues involved in the management of intellectual property in biotechnology. This would facilitate the development of greater public/private sector collaboration in biotechnology.

4. Negotiating skills

Improved skills within the IARCs, NARSs and the international development agencies are needed to negotiate (where requested) on behalf of consortia of developing countries, to obtain access to new technologies likely to be useful when adapted to commodities important in the Third World.

5. Patenting of IARC inventions

The IARCs need to identify the advantages and disadvantages of patenting their significant inventions in biotechnology. The ability of IARCs to identify such advantages will be critical for the negotiation of substantive collaboration with private companies in biotechnology.

Chapter nine:
International agricultural research centres

The IARCs ... will need to develop substantive biotechnology
programs within the next decade. This will require ... additional,
targeted funds.

Introduction

Most of the international agricultural research centres (IARCs) that deal
with crop or livestock improvement are already involved in biotechnology
research, at either modest or substantial levels. Their current programs are
described in detail by Plucknett *et al.* (1990). Some potential applications
of biotechnology in the crop improvement programs of the centres are
listed in Table 9.1.

IARCs can play an important role in adapting the tools of modern
biotechnology to the needs of Third World agriculture. In order to play this
role to its full potential, there will need to be further shifts in research
strategies and the reallocation of resources and personnel in some centres.
The major clients of the centres' efforts in biotechnology are likely to be
the many smaller and poorer countries (70 or more), rather than the more
technologically advanced countries (ten or less). The latter are developing
their own capacity in biotechnology and require little assistance from the
international centres, except in an occasional advisory role.

In order to be a serious player in biotechnology, an international centre
needs to have:

1. A critical mass of in-house scientific expertise to monitor, choose and
utilise new products and processes, and adapt them to the real limiting
factors in Third World agriculture.
2. Skills to acquire new technologies from the public or private sector in
industrialised countries, under suitable licensing and/or royalty arrangements if necessary.
3. Scientific and managerial skills to develop well-chosen and effective
collaborative research programs with the best public and private-sector
laboratories in industrialised and newly industrialised countries. This will
require legal and financial skills not commonly available at the centres, and

which may be able to be better provided on a CGIAR system-wide basis.

4. Research capacity to integrate new technologies into existing R and D programs (particularly plant breeding).

5. Support for the necessary complementary research in physiology, pathology, entomology, cytogenetics, etc. to provide the scientific basis for future genetic improvements.

6. Adequate biosafety guidelines, compatible with the regulatory environment of the host country.

Some centres have already moved their research programs in new directions, while others are in the planning process. The IARCs presently invest approximately US$12 million per year in modern biotechnology. Approximately US$8 million of this is for livestock biotechnology at ILRAD in Kenya. *If the IARCs are to continue to be centres of excellence in tropical agricultural science, most will need to develop substantive biotechnology programs within the next decade. This will require the reallocation of some existing resources and personnel, and the provision of additional targeted funds.*

Bilateral and multilateral development agencies and potential public-and/or private-sector collaborators will have the opportunity to support individual segments of biotechnology activity at the IARCs in the areas closest to their particular interests. Structuring support in such a manner will assist centres with the costs and personnel associated with integrating new technologies into conventional R and D programs.

Early applications of new targeted funds could be to: (1) facilitate the application of new biotechnologies (such as RFLP mapping) to plant breeding programs; (2) support the application of modern biotechnology to integrated pest management (including biological control) with the aim of reducing pesticide usage; and (3) introduce novel means of plant disease resistance especially against the economically important virus diseases.

Technology acquisition

In order for the IARCs to adapt the tools of biotechnology to the demands of Third World agriculture, new methods for acquiring technologies need to be considered. The centres need to be able to negotiate in a creative fashion with the private sector on their own behalf and, if so required, on behalf of consortia of developing countries. This may involve licensing a particular technology for use by a centre, in a manner analogous to the purchase of propriety computer software. Identifying and understanding the relative strengths of the IARCs and the private sector is the first step in defining working relationships between the two parties (USAID, 1989; James and Persley, 1990).

To acquire new technologies, especially in an area that is changing as

Table 9.1. Potential applications of biotechnology for integration with conventional cultivar development.

Components of conventional germplasm-based technologies	Conventional time-span	Potential biotechnology contributions
Germplasm		
A. Acquisition exchange	1 year	*In vitro* culture, disease indexing and eradication, micropropagation
B. Conservation	Ongoing	*In vitro* conservation, gene libraries
C. Evaluation	2 seasons	Molecular diagnostics, RFLPs
D. Germplasm enhancement	3–5 seasons	Embryo rescue, molecular diagnostics, selection in tissue culture, somaclonal variation, gene transfer
E. Wide hybridisation	2 years	Embryo rescue, somaclonal variation, anther culture, protoplast fusion
Breeding		
A. Selection of parental germplasm		Molecular diagnostics, tissue culture derived lines, gene transfer
1. Elite lines		
2. Adapted populations		
3. Exotic materials		
B. Initial development cross (F_1)	1 year	
C. Production and selection of segregating lines (F_2–F_3)	2 seasons	Somaclonal variation, anther culture, molecular diagnostics, RFLP mapping
D. Controlled inbreeding (F_4–F_7)	3–4 seasons	
E. Bulk increase of finished lines	2 seasons	Pathogen elimination, micropropagation
Testing		
A. Observational trials and/or preliminary testing	2 seasons	
B. International trials	2 seasons	Molecular diagnostics
C. Advanced testing in national co-ordinated trials	2 seasons	
D. Farmer's field trials	2 seasons	
Distribution		
A. Bulk increase	1–2 seasons	Micropropagation
B. Certification	1 season	Disease indexing and eradication
C. Quarantine	1 season	Disease indexing, molecular diagnostics, micropropagation

Source: Plucknett *et al.* (1990).

rapidly as biotechnology, the centres need the capability to observe, choose and utilise ideas that might prove useful in crop or livestock improvement. For most centres, this will require having scientists on their staff with the capability to keep up with scientific advances, understand them, and to judge whether the advances will be useful in agricultural research. To do this well, such scientists must be able to span the gap between basic biological sciences and agricultural sciences. *More bridging courses in modern biotechnology may be useful for senior centre scientists and research managers to provide more familiarity with the new technologies available, and their potential contribution to their current R and D programs.* ILRAD has used this approach, with some success, to increase its critical mass of modern scientific expertise.

International centres located in countries remote from new developments in modern biology may wish to consider innovative staffing arrangements to facilitate their access to new scientific developments. For example, a centre staff member could be located in an advanced institute in an industrialised country, with a brief to monitor new developments in modern biotechnology and establish collaborative linkages to the centre's R and D programs. Outposting staff in this way may prove more profitable than trying to ensure that a biotechnologist located at centre headquarters keeps abreast of such a rapidly changing field.

The acquisition of desirable biotechnologies necessitates access to new expertise at the international centres. Decisions must be made as to what type of technology is needed, to discriminate between the various options available, expected costs ascribed to each technology and the expected time required to obtain practical applications for agriculture. Once this decision-making process is in place and a technology selected, the means for its acquisition must be decided (Plucknett *et al.*, 1990).

Acquiring new technologies should not be limited to those of the public domain alone. Recent advances in the private sector, coupled with their interest and ability in working with the IARCs, should be carefully considered as an option. Collaborative relationships can be effective mechanisms for acquiring and developing these new technologies. Technology acquisition, development and application by the IARCs and the national agricultural research systems (NARSs) could benefit greatly from expanded linkages between the centres, national institutions and biotechnology practitioners elsewhere, both public and private.

A brief comparison of the main attributes of the Green Revolution and their relation to the current era of new biotechnologies is given in Table 9.2. In the light of the different attributes of biotechnology, the IARCs need to consider: (1) distributing improved germplasm through the private sector as well as through national research systems; (2) moving their research upstream towards more strategic programs; (3) relying on patents or other property rights for processes and products, and obtaining such protection of proprietary lines prior to use; (4) integrating molecular and cellular biology with conventional breeding to achieve new advances; (5) complying with national and international regulatory standards, especially those involving genetic engineering; (6) maintaining and enhancing strategic alliances for collaborative research required to augment centre-oriented research; (7) undertaking more direct commercial involvement as national programs and IARCs collaborate with the private sector; and (8) depending upon new germplasm to achieve genetic gains (Plucknett *et al.*, 1990).

Table 9.2. Attributes of the Green Revolution and their relation to modern biotechnology.

Green Revolution attribute	Potential impact of new advances
Improved germplasm distributed through NARS	Distributing improved germplasm through NARS as well as through private sector
CGIAR centres primarily conducting downstream or applied and adaptive research	Encouraging CGIAR centres to move research upstream towards more strategic programs
Patents and plant variety protection used minimally	Research institutes using patents for product protection of proprietary lines prior to use
Considered conventional research technologies sufficient to sustain agronomic progress	Integrating molecular and cellular biology with conventional breeding for agronomic advances
Sought minimal involvement with regulatory concerns regarding biological manipulation and transport, except for quarantine	Complying with national and international regulatory standards, especially those involving genetic engineering
Maintained strong reliance on research based at the IARC for breeding advancements	Maintaining and enhancing collaborative research required to augment centre-based research
Required little commercial involvement for technology dissemination	Considering more direct commercial involvement
Depended upon germplasm to achieve genetic gains	Using new sources of germplasm to increase productivity through genetic manipulations

Source: Plucknett *et al.* (1990).

Future directions

Integration of research and development

Four segments of conventional plant variety development ((1) germ-plasm, (2) breeding, (3) testing and (4) distribution) are presented in Table 9.1. Consideration of each segment and its ability to incorporate new technology generates greater understanding of where technologies are most needed. Strategies which advance agricultural development must begin with an effective conventional research program. This implies a commitment to a sound research and development infrastructure, including the following three broad categories of technology: (1) conventional technologies (i.e. plant breeding, animal husbandry); (2) cell and tissue culture; and (3) molecular biology and biochemistry. Such a strategy will then provide the framework for effective integration at the research level. It can best be done through interdisciplinary teams which may include external representatives from the private sector for market analysis, and from legal authorities (Plucknett *et al.*, 1990).

An organisational strategy must also be put into place to provide for distribution, marketing and product improvement. Strategic alliances should be formed as part of this organisational strategy, using linkage through collaboration or more formal joint ventures and partnerships. Opportunities to commercialise technologies, and hence benefit from revenue-generating ability, should be considered and employed whenever possible. Research at the IARCs and NARSs should take advantage of targeting a portion of their research portfolio towards commercial opportunities.

Collaboration with advanced laboratories

Some of the collaborative arrangements developed by the international centres and advanced laboratories in industrialised countries are listed in Table 9.3. ILRAD and CIP have established particularly strong collaborative linkages, with both public and private-sector laboratories. Such collaborations are essential for centres if they wish to keep abreast of, and involved in, modern biology.

Many of the international centres can play or are playing a central role in advanced networks that link centres with institutions developing new technologies. Some examples of existing biotechnology networks are given in Table 9.4. Such networks play an important role in targeting research, developing methodologies, establishing new research channels, expanding awareness of centre-based research and providing a means to acquire new

Table 9.3.   Current applications of biotechnology for commodity crops of some IARCs.

Commodity crop	Biotechnology activity	Centre and/or collaborators	Conventional component
Banana/plantain	*Embryo rescue* *In vitro conservation* *Micropropagation* *Somaclonal variation*	IBPGR/IITA/INIBAP/ CIRAD, France/ University of Leuvan	Breeding Germplasm Distribution Breeding
Cassava	*Anther/microspore culture* Achieving homozygosity to express recessive traits	CIAT	Breeding
	Electrophoresis Germplasm characterisation *in situ* and *ex situ*	CIAT and University of Manitoba	Germplasm
	In vitro conservation Clonal depository Pilot *in vitro* gene bank	CIAT CIAT/IBPGR	Germplasm Germplasm
	Somatic cell culture Regeneration through embryogenesis from leaf cell culture	CIAT, IITA	Breeding
	Shoot-tip culture/Heat treatment Elimination of viral and bacterial diseases	CIAT	Distribution
	Transformation Insertion of synthetic genes to increase selected amino acid production	CIAT/Louisiana State University (LSU)	Breeding: value-added
	Insertion of viral coat protein to confer virus resistance	CIAT/IITA/ORSTOM and Washington State University	Breeding
Common beans	*Electrophoresis* Germplasm characterisation using phaseolin and isozymes	CIAT	Germplasm
	Gene pool relations and evolution	CIAT/University of California, Davis	Germplasm

	Screening resistance to bruchid insects	CIAT	Breeding
	Embryo rescue		
	P. vulgaris × P. acutifolius crosses	CIAT	Germplasm
	RFLP mapping		
	Resistance to bacterial blight	CIAT/University of Florida	Breeding
	Somatic cell culture		
	Plant regeneration from cell suspensions	CIAT/Tissue Culture for Crops Project, USA	Breeding
	Plant regeneration from callus culture	CIAT/Tissue Culture for Crops Project, USA	Breeding
Cowpea	*Resistance to insects*	IITA/University of Napoli Purdue University	Breeding
	Monoclonal antibodies		
	Virus strain detection	IITA/Canada Agriculture	Testing
	Somaclonal variation		
	In vitro screening for stress tolerance to aluminium and cold	IITA/University of Napoli	Breeding
	Transformation		
	Insertion of genes for resistance breeding	IITA/University of Napoli Purdue University	Breeding
Food legumes and oilseeds	*Electrophoresis*		
	Characterisation of storage proteins in wild relatives of chickpea, lentils	ICARDA	Germplasm
	Elisa		
	Detection of peanut viruses, aflatoxin B	ICRISAT	Testing
	Embryo rescue		
	Wide hybridisation in groundnut, chickpea, pigeonpea	ICRISAT	Germplasm
	Cowpea by wild *Vigna* species	IITA	Germplasm
	Phaseolus by wild species	CIAT/AGCD	Germplasm

Table 9.3. cont.

Commodity crop	Biotechnology activity	Centre and/or collaborators	Conventional component
Potato	*Anther culture* Haploid plantlets from pollen and producing homozygous tetraploids	CIP/ENEA, Italy	Distribution
	Electrophoresis Phanerograms to verify duplicate accessions in potato genebank	CIP/University of Braunschweig, FRG	Germplasm
	In vitro plantlets Induction of *in vitro* plantlets	CIP	Distribution
	Meristem culture/Heat therapy Virus/viroid elimination	CIP	Distribution
	In vitro use of chemotherapy and thermotherapy	CIP/Cornell University/DARA, Victoria	Distribution
	Monoclonal antibodies Detection of viruses Y,A, and leafroll	CIP/Swiss Federal Agriculture Research Station	Testing
	Nucleic acid diagnostic probes Potato spindle tuber viroid	CIP/USDA	Testing
	cDNA probes for virus detection	CIP/North Carolina State University	Testing
	Protoplast fusion Transfer mithochondrial-coded male sterility traits	CIP/Weitzman Institute of Science, Israel	Breeding
	RFLP mapping Construction of molecular map for use in back-crossing	CIP/Cornell University	Breeding
	Somaclonal variation *In vitro* selection for salt and drought tolerance in wide crosses	CIP/ENEA, Italy	Germplasm

	Transformation		
	Insertion of synthetic genes to increase tuber amino acid production	CIP/Louisiana State University	Breeding
	Resistance to leafroll virus and PSTV through antisense constructs	CIP/Louisiana State University	Breeding
	Resistance to fungal and bacterial diseases through transformation	CIP/Louisiana State University	Breeding
Rice	*Anther culture*		
	Selection of stress-tolerant lines	IRRI	Breeding
	Rapid achievement of homozygous lines from F_1 sexual crosses	IRRI	Breeding
	Homozygous diploids for testing in the southern cone of South America	CIAT	Testing
	Embryo rescue		
	Resistance to brown planthopper	IRRI	Testing
	Monoclonal antibodies		
	Diagnostics for rice tungro and rice grassy stunt viruses	IRRI	Testing
	Nucleic acid diagnostic probes		
	Rice blast and bacterial blight	IRRI	Germplasm
	Protoplast fusion		
	Hybrid production between species with incompatibility barriers	IRRI	Breeding
	RFLP mapping		
	Nuclear and cytoplasmic variation determined in breeding material	IRRI	Breeding
	Somaclonal variation		
	Selection for tolerance to salt and aluminium	IRRI	Breeding
Sweet potato	*Embryo rescue*		
	Sweet potato cultivars by wild *Ipomoea* species for introgression	CIP	Germplasm

Table 9.3. cont.

Commodity crop	Biotechnology activity	Centre and/or collaborators	Conventional component
	In vitro conservation		
	Clonal depository for over 1500 clones	IITA	Germplasm: conservation
	Monoclonal antibodies		
	Index sweet potato viruses disease	CIP/USDA/ARS	Distribution
	Nucleic acid diagnostic probes		
	Detection and characterisation of sweet potato viruses	CIP/North Carolina State University	Testing
Wheat	*Electrophoresis*		
	Provide biochemical markers for alien germplasm in wide crosses	CIMMYT	Germplasm
	Monoclonal antibodies		
	Detection of barley yellow dwarf virus	CIMMYT	Testing
	Somaclonal variation		
	Somaclone selection from spring wheat for tolerance to salinity	CIMMYT/University of Colorado Tissue Culture for Crops Project (TCCP), USA	Breeding
	Regeneration and somaclone selection in durum bread wheats	CIMMYT/TCCP	Breeding
	Tissue culture for alien gene-introgression		
	Use of *Aegilops* as source of karnal bunt resistance	CIMMYT and TCCP	Germplasm: prebreeding
	Use of *Agropyron* for tolerance to *Helminthosporium*	CIMMYT and TCCP	Germplasm: prebreeding
	Wheat × disomic addition lines containing rye chromosomes	CIMMYT and TCCP	Germplasm: prebreeding

Source: Plucknett *et al.* (1990).

technologies. The centres gain because they can link their own programs with those of advanced institutions on problems of mutual interest. Collaborating institutions benefit because they can learn more about the genetic resource collections held by the centres as well as major agricultural problems that might benefit from biotechnology approaches. Such network warrant further support from international development agencies which could fund the involvement of additional collaborators from the NARS and laboratories in industrialised countries.

Currently, the most well-established networks related to the work of the IARCs is the Rockefeller Foundation's network on biotechnology in rice. Here advanced laboratories in Europe, USA and elsewhere are linked in efforts with the International Rice Research Institute (IRRI) to improve rice production by overcoming certain major constraints using biotechnology. The network collaborators expect to have new rice varieties emanating from their work in the field by the mid 1990s – an impressive achievement.

In September 1988 a group of scientists met at CIAT to establish the 'Advanced Cassava Research Network'. The network's primary objective is to integrate new biotechnology approaches into research efforts to overcome production constraints in cassava. Three criteria were established for determining where research efforts should be placed: (1) the problem should be one where emerging technologies show promise; (2) conventional research efforts have not been fully successful; and (3) solutions should have promise for widespread application. Targets which fit these criteria include: low cyanide, virus disease resistance and true seed production of cassava (see Chapter 2 and Bertram, 1990).

These emerging networks provide a means to link priority research areas at the IARCs with new technologies that might otherwise not be directed at their mandate crops. These international efforts should help mount additional research on 'orphan commodities' important to the Third World, although they may not be the targets for modern biotechnology in the industrialised countries.

Table 9.4. Biotechnology network[a] applying new technologies to Third World agriculture.

Network	Sponsors	Major activities
Rice Molecular Biology	Rockefeller Foundation	1. Application of molecular biology of rice
International Plant Bio-technology Network	Tissue Culture for Crops Project (TCCP): Colorado State University; USA	1. Sponsor international plant biotechnology conferences 2. Publish directory of tissue culture practitioners in developing and industrialised countries 3. Provide short- and long-term training in cell culture 4. Publish newsletter
Anaplasmosis/Babesiosis Network	Improved Animal Vaccines through Biotech; Washington State University; USAID	1. Compile and distribute research bibliography on anaplasmosis and babesiosis 2. Quarterly ABN newsletter 3. Sponsor international meetings.
Cassava Transformation	ORSTOM, Washington University (St Louis), Rockefeller Foundation	1. Transformation and regeneration of cassava 2. Insertion of genes for virus resistance
Advanced Cassava Research Network	CIAT/IITA	1. Apply new technology to research constraints of cassava 2. Link IARCs with biotechnology labs in US and Europe 3. Determine research priorities for application of biotechnology to cassava

Note: [a]The list of networks given is illustrative, rather than comprehensive.
Source: Cohen (1989).

Findings

Additional support should be provided to the IARCs to enable the expansion of the following types of activities:

1. Scientific expertise at the IARCs

Facilitate interchanges, including university and industrial postdoctoral research; development of interdisciplinary teams to handle new technologies and apply them to specific agricultural constraints; out-posting of centre staff to advanced laboratories, to monitor new developments and identify potentially useful technologies.

2. Technology acquisition

Facilitate the acquisition of new technologies, including ones from the private sector, through purchase, licensing, or other agreements on royalties; collaborative research designed to benefit interests of both parties; market analysis; technology appraisal.

3. Verification

Enable testing of model systems in contained experiments prior to wider adoption.

4. Application

Encourage participation of national programs and commercial concerns in the application of biotechnology to agricultural development.

5. Outreach

Provide opportunities for training, extension and adaptive research with NARSs.

6. Policy and management issues

Collaborate with NARSs on the definition of national policies on biotechnology and the preparation of national programs appropriate to the size and scientific infrastructure of a particular country.

Chapter ten:
World Bank investments

The time is now right to review ... the Bank's current investments in
the agriculture sector ... to maximise the benefits derived by
NARSs ...

Introduction

Biotechnology is offering innovative possibilities for increasing crop and
livestock production and for the protection of the environment by the
reduced use of agrichemicals. The major thrust is presently directed
towards medicine, industry and agriculture in the industrialised countries,
with significant investments by transnational companies. As a major
supporter of agricultural research, the World Bank should ensure that the
needs of the Third World are not ignored in the current research and
development efforts in biotechnology (Pritchard, 1990).

The World Bank provides support for agricultural research primarily by
loans for national agricultural research projects; loans for rural development projects that contain a research component; and grants to the IARCs.
There are also a few loans in the education and science and technology
area which contain an agricultural research component.

The most important constraints limiting the development of strong
NARSs are: (1) shortages of well-trained scientific staff; (2) lack of
government commitment to research as exemplified by budgetary cuts and
low levels of funding for operational expenses; (3) low salary levels for
research staff and lack of adequate personnel management procedures and
policies; and (4) lack of well-established research/extension linkages
(Pritchard, 1990).

Bank support for agricultural research

In the period 1981–7 the World Bank invested approximately US$575
million in 21 free-standing national agricultural research projects, by means
of loans and credits (Table 10.1). In addition, there were 209 agricultural
and rural development projects with research components. The total cost
of the agricultural research elements in these rural development projects

Table 10.1.   World Bank free-standing agricultural research projects 1981–7.

Fiscal year	Country	US$ millions	
		Total project cost	Loan/credit amount
1981	Pakistan	40.1	24.0
	Thailand	91.5	30.0
	Brazil	150.1	60.0
1982	Yemen AR	32.4	6.0
	Philippines	77.0	45.0
	Peru	83.3	40.0
	Senegal	106.1	19.5
1983	Papua New Guinea	24.4	14.1
	Colombia	206.6	63.4
1984	Bangladesh	32.2	24.5
	Zimbabwe	177.1	13.1
1985	Ethiopia	32.1	22.0
	Malawi	49.9	23.8
	Rwanda	18.0	11.5
	China	59.0	25.0
1986	India	110.9	72.1
	Sudan	38.0	22.0
1987	Sri Lanka	26.5	18.6
	Zambia	38.8	13.0
	Malaysia	43.2	9.0
	Cameroon	43.2	17.8
Total		1474.5	575.0

Source: Pritchard (1990).

was about US$2.1 billion. Additional support for agricultural research was provided through 12 educational projects, 16 policy-based loans, and grants to the CGIAR-affiliated IARCs and other non-associated international centres (Pritchard, 1990). The grant funds provided to IARCs in 1988 was US$32 million.

Some of the Bank's current support for agricultural research is used for agricultural biotechnology. It is difficult to estimate the precise amount, since biotechnology is often a minor component which is included within the overall project description without any detail being provided on the funds especially allocated to biotechnology.

Bank lending for agricultural biotechnology

Biotechnology in the agricultural sector program

The World Bank's specific involvement with modern biotechnology is relatively recent, but already substantial. Lending for modern biotechnology is, however, minor when compared with lending for other purposes. Traditional biotechnology has been supported for many years through the Bank's lending programs for agriculture, education and science and technology.

Between 1982 and 1989, the Bank supported ten agricultural projects with specific biotechnology components. These projects are located in Brazil, Cyprus, Hungary, Indonesia, Malagasy, Malaysia, Rwanda, Senegal, Sri Lanka and Sudan (Table 10.2). Additional support has been provided for biotechnology through the education and science and technology sectors, for projects in Brazil, Indonesia and Portugal. The total estimated loans and credits for these projects from 1982 to 1989 are US$93.4 million (Table 10.3).

The projects include support for infrastructure, laboratory facilities and equipment for biotechnology, training and R and D programs involving,

Table 10.2. Selected World Bank lending for biotechnology in the agriculture sector program, 1982–9[a].

Year	Country	Project	Biotechnology component
1982	Cyprus	Fruit and vegetables	Tissue culture
1983	Malagasy	Cotton development	IPM[b]
1985	Rwanda	Agricultural research	IPM[b]
1985	Hungary	Crop production improvement	Biotechnology[c]
1986	Sudan	Agricultural research extension and training	IPM[b]
1987	Sri Lanka	Agricultural research	Tissue culture
1987	Malaysia	National forestry research	Tissue culture
1989	Senegal	Agricultural research project	Biotechnology[c]
1989	Brazil	Third agricultural research project	Biotechnology[c]
1989	Indonesia	Agricultural research management	Biotechnology[c]

Notes:
[a]Biotechnology components contained in a sample of agricultural sector projects.
[b]Integrated pest management.
[c]Broadly based biotechnology program.

Table 10.3. Selected World Bank lending for agricultural biotechnology, 1982–9.

		US$ millions		
Country	Project name	Project cost	Loan	Biotechnology component
Portugal	Tech. Education	77.1	32.0	28.1
Brazil	Science–technology	215.4	72.0	31.2
Indonesia	Science–technology	153.7	93.0	—
Indonesia	University Dev. II	244.5	147.0	33.7
Indonesia	Agricultural Research Management	50.4	27.5	0.4
Total		741.1	371.5	93.4

for example, tissue culture of various crop and forest species, and bioprocessing (Table 10.2).

In the 1988–9 financial year, three additional projects (in Senegal, Brazil and Indonesia), which include biotechnology components were presented to the Bank for approval. In the Third Agricultural Research Project in Brazil, support will be provided for both modern and traditional biotechnology through the financing of new laboratory facilities, additional staff salaries and training for existing staff. Research will be conducted on cell and molecular biology, and the biological control of pests. The targets include vegetables, beans and oil palm. Animal biotechnology is to be directed towards cattle and horses.

The new Agricultural Research Management Project in Indonesia will provide support for the application of biotechnology to export crops. Research will be conducted on cell and tissue culture and on microbial and enzyme fermentation processes for crop processing. The project will also finance laboratories, equipment and vehicles.

Biotechnology in other Bank lending programs

More support has been provided for biotechnology through the educational and science and technology sectors than through agricultural sector lending. The Science and Technology Training Project in Indonesia (1985) provided 1500 overseas training fellowships, of which seven were reserved for biotechnology at the Institute of Science and Technology, and 32 at the Agency of Assessment and Application of Technology (11 tissue culture, six legal patents, 15 biotechnology).

The Second University Development Project in Indonesia (1985), aimed at the development of graduate level biotechnology programs at the Institute of Technology (Bandung), the Institute of Agriculture (Bogor) and the University of Gadjah Mada. The project finances research, laboratory facilities and equipment at these institutions and will increase the number of teaching staff with higher degrees. Of the 1666 fellowships provided, 249 are for biotechnology.

The Technical Education Project in Portugal (1987) supports the construction of a Biotechnology Institute near Lisbon and a regional laboratory near Porto to service the needs of the northern region. The institute will conduct research on biotechnology relevant to agriculture, aquaculture, pharmaceutical products and food processing. Besides civil works, the project will finance the equipment necessary for high-grade laboratories, library facilities and 236 months of technical assistance. The institute would be responsible for co-ordinating biotechnology research and would encourage collaboration with private industry in biotechnology research and development.

The Project for Science and Technology in Brazil (1985) provides finance for subprojects which are appraised by Brazilian intermediary agencies. These include biotechnology subprojects which aim to build a basic capacity for biotechnology in Brazil by developing expertise in the biological sciences and related disciplines. Under the project, available resources will be concentrated on priority technical problems and inter-actions encouraged between universities and the emerging industrial operations. In agriculture, subprojects are included on developing plants resistant to adverse conditions; increased efficiency in the use of crop nutrients; cell and plant tissue culture; atmospheric nitrogen fixation; and animal vaccines.

Emerging trends

With such recent involvement in biotechnology it is possible to draw only preliminary conclusions from the World Bank's current lending program for biotechnology. Until the projects have been evaluated it is also difficult to assess accurately the amount of financial support the Bank is providing for biotechnology because: (1) the various components dealing with different aspects of biotechnology usually are not costed separately; (2) the content of the research programs and the research priorities are often determined during project implementation and are not described in the appraisal report; and (3) where the complete national research program is supported, the individual topics and disciplines are not outlined in a manner which enables biotechnology programs to be clearly identified (Pritchard, 1990).

It is possible to determine some underlying characteristics of the current World Bank lending for biotechnology: (1) that only the more econ-omically advanced countries have received relatively substantial support for biotechnology, and this has concentrated on training; (2) the need for a strong and effective association between the public sector and the private sector is recognised in most projects; and (3) the Bank's current emphasis in biotechnology is on pharmaceuticals and human health rather than agricultural biotechnology (Pritchard, 1990).

Relationship between the World Bank and the CGIAR centres

The World Bank is a substantial supporter of the CGIAR. It is one of the three co-sponsors (with the United Nations Development Programme (UNDP) and the Food and Agriculture Organization (FAO)); it provides the Chairperson for the Group, and hosts the Secretariat; it also provides approximately 15% of the total budget of the CGIAR-supported institutes as unrestricted core funds (a grant of US$30 million in 1988).

In spite of the substantial support it provides to the IARCs, the Bank has been satisfied to accept a largely passive role within the CGIAR system in terms of the research directions of the centres. It has not attempted to influence the programs of the centres by preferential funding nor link the IARCs directly with other Bank-supported activities. Where centres have been involved in Bank-supported national agricultural research projects, it has been at the request of the countries themselves rather than through Bank initiative (Pritchard, 1990).

The World Bank is, therefore, in a paradoxical situation with regard to the course of agricultural research in the centres. On the one hand, it is a large financier of the CGIAR system and a large consumer of its output of research results; yet, on the other hand, it exercises only limited influence on the research of the centres and the balance of activities conducted at the centres. *As the Bank is increasing its investment in agricultural research and in the agricultural sector generally, the time is now right to review these activities and consider ways through which the Bank's current investments in the agriculture sector could be better linked, to ensure the efficient use of the available resources for developing new technology, and to maximise the benefits derived by NARSs* (Pritchard, 1990).

Findings

There are different instruments within the World Bank that could be used to provide additional support for the application of biotechnology to agricultural development. Possible mechanisms are:

1. Loans

Additional loans and credits for national agricultural research projects in individual countries could usefully have more clearly defined biotechnology components. These could include a new form of technical assistance featuring research partnerships between scientists in developing countries and scientists in advanced laboratories elsewhere. This would facilitate the acquisition and adaptation of new technologies to the needs of agricultural development in the Third World.

2. Joint Ventures

The International Finance Corporation (IFC) assists the private sector in developing countries by providing capital (including venture capital) and management expertise. This instrument could be used to facilitate the participation of local private-sector companies in the commercial development of biotechnology, particularly for larger projects that have a biotechnology component. The IFC may also be able to facilitate the acquisition and adaptation of new technologies by its member countries.

3. Grant funds

The Bank could provide additional targeted grant funds to the IARCs to expand their capacity in modern biotechnology and their ability to undertake additional collaborative research with advanced laboratories.

Chapter eleven:
Options for international development agencies

New initiatives rather than new institutions are required, and these must build on the traditional strengths in agricultural research, not displace them.

Introduction

The challenge facing agriculture in the Third World is to double food production in the next 25 years. More and better grain varieties, fertiliser and irrigation will continue to provide the basis of the production response needed. The constraints of energy prices, environmental deterioration and production plateaux are now more binding and new innovations are being sought increasingly through the application of biotechnology.

Mobilisation of the limited resources available to NARSs will require new policy and institutional arrangements, which seek to develop stronger collaborative research and training linkages with established biotechnology laboratories elsewhere, and encourage greater strength in private-sector research in agriculture.

The future access to advances in science and technology, hitherto freely available as public goods, seems likely to be constrained further as intellectual property rights and private-sector research assume greater importance in the industrialised countries. *An ability to identify technology developments in advanced laboratories relevant to local problems, and to negotiate the transfer and adoption of this technology, will be attributes of great importance in research management in the future* (Javier, 1990).

The biotechnological innovations already successful elsewhere (for example, micropropagation and pathogen elimination in cell and tissue culture, improved diagnostic techniques and vaccine production for local-ised diseases) are likely to be the early targets on which to focus initial efforts in the Third World.

Significant differences in the manner in which biotechnology research is organised in different countries are apparent. Some countries have chosen to concentrate their resources in a central biotechnology institute. Others prefer to integrate biotechnology into their existing agricultural research institutes. Whichever institutional mechanism is chosen, the critical linkage is to establish effective interdisciplinary collaboration amongst the molecular biologists, pathologists, agronomists and breeders. *New initiatives rather than new institutions are required, and these must build on the traditional strengths in agricultural research, not displace them.*

How best to integrate biotechnology into existing agricultural research programs, how to plan the training and retraining of staff, how to access and adapt technology developed elsewhere, how to arrange and finance collaborative agreements with other biotechnology enterprises, how to protect the rights to unique germplasm and how to ensure adequate environmental and social safeguards are critical questions that recur in all countries seeking to develop a capability in biotechnology.

A first step in providing the answers required centres on building a capability in the basic biological sciences. Without this no country can reasonably aspire to the productive use of biotechnology. At present relatively few countries have this strength. Much of the skill in applying biotechnology to agricultural production problems resides in the private sector of the industrialised countries. These private-sector skills are complemented in these countries by a strong and publicly financed research system which, in turn, is reinforced by the demands made upon it by an active agricultural industry. *Strengthening capability in biotechnology depends upon the development of collaborative research agreements between the public and private sectors across country boundaries. Facilitating the transfer of relevant information and techniques is likely to be one of the most important roles the international development agencies can play in helping countries capture the benefits this new technology offers.*

A biotechnology policy that seeks to increase the flow of benefits from biotechnology in countries with a strong research capacity to those much weaker in this respect would have three major objectives: (1) the need to ensure the benefits are available to all developing countries; (2) the need to encourage research and development activities which strengthens the productive capacity of small farmers; and (3) the need to protect the environment by the judicious use of the new technologies.

The translation of these objectives into an operational response by the international development agencies hinges upon a knowledge of the technology available; of finding ways to transfer that information to target problems along with the skills to use it; of negotiating the licensing of technology, the use of which is governed by intellectual property rights; and of ensuring that adequate safeguards are in place to protect society from any unwise release of new organisms.

These tasks go well beyond the technical responsibilities in project preparation that international development agencies presently accept. They are skills also not yet consolidated in the nonprofit foundations, the IARCs, or in private consultant groups. This is a major gap limiting much greater support to agricultural biotechnology by the international development agencies. It is a finding these agencies now need to address, by strengthening their in-house capacity in biotechnology, and by accessing appropriate sources of external expertise to assist them in project preparation.

Innovative funding mechanisms are required to enable international development agencies to work more closely with the private companies (both local companies and transnational companies). Mutual agreement would need to be reached on a case-by-case basis on the management of intellectual property, including the use of any proprietary technologies involved.

Many international development agencies work extensively with private companies in other sectors (e.g. engineering), and could draw on this expertise while negotiating mutually acceptable arrangements with individual companies in biotechnology.

Orphan commodities

An 'orphan commodity' is defined here as a commodity in which there is likely to be little or no investment in modern biotechnology in industrialised countries, either because the commodity is not important in temperate areas or because there are no likely profits for transnational companies.

An orphan commodity is not necessarily a small commodity. For example, banana/plantain, coconut and cassava would fit the above definition of an orphan commodity, although all are widely cultivated crops in the Third World. Another group of orphan commodities is the small, specialised, but potentially high-value crops which could benefit from the application of modern biotechnology (e.g. tropical fruits, spices, vegetables).

Special attention should be paid to orphan commodities by the international development agencies in order to facilitate the application of modern biotechnology to the problems peculiar to these crops. Different funding mechanisms are possible for the international development agencies to support research and development on different types of orphan commodities. Some possible mechanisms are listed in Table 11.1.

It may also be useful to establish a new funding mechanism, under international auspices, for an 'orphan commodities program'. This would provide support for research on orphan commodities by public- and private-sector institutions in industrialised and developing countries (Persley, 1990c). Priority could be given to involving private-sector research laboratories in contract work on commodities in which they are not presently interested, but for which they have relevant technologies. An orphan commodities program would be analogous to the 'orphan drugs program,' presently sponsored by the World Health Organization, the World Bank

Table 11.1. Possible funding mechanisms to support the applications of modern biotechnology to orphan commodities.

Orphan commodity	Funding mechanism
IARC-mandated orphan commodities (e.g. cassava)	Grants to IARCs (including new targeted funds for biotechnology)
Large orphan commodities (e.g. banana/coffee)	Soft loans to major producing countries, on commodities where short- to medium-term applications of biotechnology are possible
Small orphan commodities (e.g. tropical fruits)	Grants (bilateral/regional)

and some philanthropic foundations to sponsor the development of new pharmaceuticals for Third World markets. Some of the major pharmaceutical companies are participants in this program, as are several public-sector institutions in Europe and the USA.

Biotechnology advice and information

There is a need for individual countries to have ready access to impartial advice as to how best to integrate modern biology (biotechnology) into existing research efforts and agricultural development programs. This advice should be available prior to decisions on whether to seek substantial external financing of new programs. Information provided needs to focus on ways to identify current problems where biotechnology may be helpful, appropriate institutional and management arrangements and the necessary policy issues, including regulatory requirements and intellectual property management. International agencies such as ISNAR may need to incorporate these services within their advice to NARSs. Early warning also needs to be available on any potential negative effects of biotechnology, either on individual commodities or countries.

There is also a need for international development agencies such as the World Bank and other bilateral and multilateral agencies to strengthen their in-house capacity to appraise biotechnology requests. This should include ensuring existing staff have ready access to current information and sources of expertise on new technology developments and the policy and management issues related to biotechnology. These sources of expertise are likely to be different to traditional consultant sources.

Findings

International development agencies: policy options

There are three main policy options for international development agencies interested in facilitating the application of biotechnology to agriculture in the Third World. These are:

1. *Public sector support* which would provide support for biotechnology in association with other support for agricultural research at public-sector institutions, including the provision of training opportunities and research facilities. This would require little change from present policy and practice in support for agricultural research.

2. *Public/private sector support* which would provide support for biotechnology at public-sector institutions within the NARSs and at the IARCs, including support for collaborative activities with advanced laboratories, in either the public or private sector.

Private companies may choose to participate in such collaborative activities either on a contract research basis, or because participation in the project may confer commercial advantages that are not linked directly to the project itself.

3. *Commercial joint ventures* which would facilitate establishment of commercially viable joint ventures between public- and private-sector agencies, in both industrialised and developing countries. These new partnerships could facilitate both the development of new products and processes, and the business systems necessary for the delivery of novel products to the end-users in the Third World.

Orphan commodities

The early applications of modern biotechnology are likely to be on crops which are primarily of interest to industrialised countries.

There is a need for additional investments in biotechnology on orphan commodities, these being ones important in the Third World but for which there are presently minimal investments in modern biotechnology, because of their lack of importance in the industrial world.

An *orphan commodities program* could be established under international auspices to sponsor collaborative research amongst public- and private-sector organisations in order to ensure that the orphan commodities also benefit from the application of new technologies. This could be analogous to an existing orphan drugs program that sponsors the preparation of pharmaceuticals for the Third World.

Biotechnology advice and information

1. International development agencies could facilitate access by developing countries to information and advice on how best to integrate biotechnology into agricultural R and D programs. These services need to focus on
problem identification;
"technology architecture", i.e. the design of ways to build and/or adapt new biotechnologies suited to the needs of individual countries;
institutional and management arrangements;
regulatory issues;
intellectual property management;
ways to facilitate access to potentially beneficial new technologies;
early warning systems for any potential negative effects, either on individual commodities or countries.

2. The international development agencies also need to strengthen their in-house capacity in biotechnology, and ensure that their professional staff have access to current sources of information and expertise in modern biotechnology. These sources are likely to be different to traditional consultant sources.

Chapter twelve:
Summary and conclusions

Modern biotechnologists or genetic engineers are becoming the new
partners of agricultural scientists.

Introduction

Modern biotechnologists or genetic engineers are becoming the new
partners of agricultural scientists. They see new ways of tackling old
problems, through the application of biotechnologies ranging from the
long-established, commercial use of microbes and other living organisms to
strategic research on genetic engineering of plants and animals. Biotechnology may hold the key to providing sufficient food for the additional
billions beyond the year 2000, and to protect and preserve the world's
genetic diversity in microorganisms, plants and animals.

Socioeconomic impact

In many countries the scope for improving agricultural output with existing
technologies is still large. However, there are limits to what can be achieved.
Environmental considerations are also checking rapid agricultural expansion, since further reductions in the area of land for pastures and forests are
unacceptable in many countries. The development of new technologies will
be one of the key factors necessary to obtain the substantial increases in
food production required to meet expanding populations in the Third
World.

The major change in agricultural research in industrialised countries in
the past decade has been the substantially increased role of the private
sector, largely in funding research in modern biotechnology. The early
applications of modern biotechnology are likely to be on commodities such
as hybrid maize that are primarily of interest to industrialised countries. In
1987, total R and D expenditure on agricultural biotechnology was
estimated at US$900 million, of which US$550 million was in the private
sector.

Significant economic impact of modern biotechnology on agricultural
production is not expected before the year 2000. After that, it will be an

increasingly important component of new technologies for crop and livestock production. This time-frame is similar to that for conventional research, in that it takes about ten years to develop a new plant variety, animal vaccine or pharmaceutical.

Biotechnology is likely to change the comparative advantages between countries and between commodities, particularly for export commodities. The application of new technologies to export commodities will improve their competitive position in the international marketplace.

The likely socioeconomic effects of biotechnology in the Third World are positive in terms of increasing productivity of tropical commodities to meet future food needs in developing countries; opening up new opportunities for the use of marginal lands; and reducing use of agrichemicals. They are also potentially negative, in that they offer the possibility of producing high-value products in tissue culture in industrialised countries, thus displacing crops presently grown for export in the Third World. Early warning systems should be developed whereby the potential negative substitution effects could be monitored and strategic adjustments recommended where economically damaging substitution effects are identified.

National strategies

In considering the approaches to agricultural biotechnology taken by certain countries, it is evident that some have well-formulated biotechnology policies, national programs and adequate financial resources. Others, although committed to biotechnology at a policy level, have not formulated a national program, and do not yet have sufficient financial support. The development of a national biotechnology strategy would be helpful in the implementation of an effective national program in biotechnology. Public-sector investments in biotechnology, and creative partnerships between public- and private-sector interests, are critical in establishing a competitive strategy in biotechnology.

Commodity analysis

The first step in assessing the usefulness of biotechnology to agriculture is to identify the problems still unresolved by conventional approaches and which may benefit from the application of new techniques. Several commodities important in the Third World have been examined to assess their current constraints to productivity, and the likely availability of new technologies to aid their resolution: Banana/plantain, cassava, cocoa, coffee, coconut, oil palm, potato, rapeseed, rice and wheat. Substantial progress may be expected in the short term (0–5 years) for potato, rapeseed and rice; in the medium term (5–10 years) for banana/plantain, cassava and coffee; and only in the long-term (10+ years) for cocoa, coconut, oil palm and wheat.

Technology assessment

Developments in biotechnology over the past decade have been accompanied by often exaggerated claims as to their likely potential impact on agriculture. More realistic predictions are now becoming available as are likely time-frames for application.

Crops

The major route for the application of new technology in crop agriculture will be through the development of new plant varieties with novel characteristics. Greatly strengthened plant breeding programs in the Third World are a necessary prerequisite for the successful application of biotechnology to crop production.

The new technologies with potential for early application in crop agriculture in the Third World are: new detection methods for pests and diseases, which could be made readily available and used widely in many countries; genetic mapping of major tropical crops, as an aid to conventional plant breeding programs; plant virus resistance, by genetic engineering of the host plant; and novel biocontrol agents for pest control to reduce pesticide use.

Microbial processing

Novel bioprocessing techniques are likely to lead to new value-added products in the food and fermentation areas.

Forestry

The likely initial applications of biotechnology in forestry are in the identification of useful genes for tree breeding and in the development of novel biocontrol agents for pests and diseases.

Livestock

The likely early applications of biotechnology in livestock production in the

Third World are in the areas of: embryo technology, especially for cattle; novel vaccines against infectious diseases; new methods to increase the efficiency of disease identification; and genetic mapping for livestock breeding programs.

Aquaculture

The likely early applications of biotechnology to aquaculture are in the development of rapid methods for disease detection, and improved methods for feed formulation and storage.

Patent issues

A major policy issue that will affect the application of biotechnology in agriculture is the management of intellectual property. The lack of patent protection is a major disincentive for private-sector investments in biotechnology in the Third World, both by local private-sector companies, and transnational corporations.

The advantages of the provision of intellectual property rights in biotechnology are to encourage the development of local research capacity, and greater in-country investments in biotechnology. The major disadvantage is that it involves giving proprietary protection to living organisms, which some consider to be part of the common heritage of mankind. Each country will weigh the benefits and costs of intellectual property rights in biotechnology, and frame its policies accordingly. Model patent systems could be developed by countries which decide to take steps to manage their intellectual property.

New mechanisms are required to facilitate access by developing countries to novel biotechnological products and processes while protecting the legitimate interests of their inventors. These mechanisms could include licensing arrangements and other contractual agreements. International development agencies and the IARCs could facilitate such access, and negotiate for the licensing of new technologies to apply to commodities important in the Third World. The IARCs could also patent their own inventions, and then license (freely, if appropriate) these inventions for use by NARSs and other collaborators.

Regulatory issues and environmental release

Another major policy issue is the regulatory climate governing the release of novel products to ensure public health and environmental safety. A safe and efficient regulatory process is in itself a comparative advantage in biotechnology.

In many countries existing legislation is sufficient to regulate the use of most agricultural products likely to be produced using biotechnology. The new requirement is for guidelines to cover the handling of genetically engineered organisms at the experimental stage, and for the assessment of any risk associated with the release of genetically engineered organisms into the environment. The new guidelines should be framed so as to complement and support existing regulatory agencies.

Several studies now report that the benefits of the use of the new technologies are likely to outweigh the risks, and that risk assessments can be undertaken to guard against the release of any potentially damaging organisms. Guidelines based on those endorsed by the OECD are now in use in several countries to allow the proper assessment of risk associated with the release of new organisms into the environment.

All countries need to establish functioning national review bodies and institutional biosafety committees, and to develop guidelines to monitor and regulate the applications of biotechnology. The IARCs also need to have their own institutional biosafety committees and ensure that these function in accordance with host-country approval mechanisms. A mechanism is required to enable national review bodies and the IARCs ready access to recent developments in risk assessment, for incorporation into their biosafety procedures. It is also desirable that biosafety reviews be conducted prior to release of any genetically engineered organisms in projects sponsored by international development agencies.

International agricultural research centres

There are substantial differences amongst the IARCs in their current involvement in modern biotechnology. Most will need to develop substantive programs within the next decade, requiring the reallocation of some resources, and the provision of additional targeted funds. It will also require legal and financial skills not commonly available at the centres and which may be better provided on a system-wide basis by the CGIAR.

World Bank investments

The options by which the World Bank could provide further support for agricultural biotechnology are: *additional loans* for specific biotechnology components within national agricultural research projects; *new joint ventures* involving the International Finance Corporation, to assist the participation of the private sector in developing countries in the commercial development of biotechnology; *additional targeted grant funds* to the IARCs to increase their capacity in biotechnology and their ability to work with NARSs in acquiring and adapting new technologies.

Options for international development agencies

There are three main policy options for international development agencies interested in facilitating the application of biotechnology to agriculture in the Third World: *Option 1. public-sector support:* provide support for biotechnology in association with other support for agricultural research at public sector institutions, including universities. This would require little change to present policy and practice. *Option 2. public/private-sector support:* provide support for biotechnology in public-sector institutions in developing countries, together with support for collaborative research with public- and/or private-sector laboratories in industrialised and newly industrialised countries. *Option 3. joint ventures:* facilitate the establishment of commercially viable joint ventures between public and/or private-sector agencies.

Orphan commodities

An orphan commodity is one on which there is little investment in modern biotechnology in industrialised countries, but which is an important commodity in the Third World (e.g. coconut; banana/plantain). Special attention should be paid to orphan commodities by international development agencies in order to facilitate the application of modern biotechnology to the problems peculiar to these crops.

Biotechnology advice and information

There is a need for individual countries to have ready access to impartial advice as to how best to integrate modern biology (biotechnology) into existing research efforts and agricultural development programs. This should be available prior to decisions on whether to seek substantial external financing of new programs. Information provided needs to focus on ways to identify current problems where biotechnology may be helpful; appropriate institutional and management arrangements; the necessary policy issues including regulatory requirements and intellectual property management; and early warning systems for any potential negative effects of biotechnology, either on individual commodities or countries. International agencies such as ISNAR may need to incorporate these services within their advice to NARSs.

There is also a need for international development agencies such as the World Bank to strengthen their in-house capacity to appraise biotechnology requests. This should include ensuring that existing staff have ready access to current information and sources of expertise on new technology developments and the policy and management issues related to biotechnology.

Conclusion

The new features of biotechnology are of sufficient importance to merit the establishment by international development agencies of innovative funding mechanisms, and new systems for the delivery of effective programs incorporating biotechnology. The design of such programs requires a policy dialogue amongst the international development agencies and individual countries, as well as the major players in modern biotechnology in the public and private sectors. It requires especially the international development agencies to devise new ways of doing business.

References

Anderson, J.R. and Herdt, R.W. (1989) The impact of new technology on foodgrain productivity to the next century. Invited paper. Conference of the International Association of Agricultural Economists, Buenos Aires, 24–31 August 1988.

Anon (1989a) *Summaries of Country Reports, May 1989.* World Bank–ISNAR–AIDAB–ACIAR Biotechnology Study Project Papers. ISNAR, The Hague.

Anon (1989b) *UK Royal Commission on Environmental Pollution* 13th Report, The Release of Genetically Engineered Organisms into the Environment, HMSO, London.

Barker, R. (1990) Socioeconomic impact. In: Persley, G.J. (ed.), *Agricultural Biotechnology: Opportunities for International Development.* CAB International, Wallingford, UK.

Barton, J. (1989) *Regulatory/Patent Issues for the Rockefeller Foundation's International Program on Rice Biotechnology.* Rockefeller Foundation, March 1989. (Available from the author, Stanford Law School, Stanford, CA 94305, USA.)

Beachy, R.N. and Fauquet, C. (1989) Abstr. Biotechnology in the control of plant virus diseases. In: *Summaries of Commissioned Papers, May 1989.* World Bank–ISNAR–AIDAB–ACIAR Biotechnology Study Project Papers. ISNAR, The Hague.

Beier, F.K., Crespi, R.S. and Straus, J. (1985) *Biotechnology and Patent Protection – An International Review.* OECD, Paris.

Beringer, J.E. (1990) Beneficial microorganisms. In: Persley, G.J. (ed.), *Agricultural Biotechnology: Opportunities for International Development.* CAB International, Wallingford, UK.

Bertram, R.B. (1990) Cassava. In: Persley, G.J. (ed.), *Agricultural Biotechnology: Opportunities for International Development.* CAB International, Wallingford, UK.

Bollinger, W.H. (1990) Commercial prospects. In: Persley, G.J. (ed.), *Agricultural Biotechnology: Opportunities for International Development.* CAB International, Wallingford, UK.

Buttel, F.H. (1990) Sociological impact. In: Persley, G.J. (ed.), *Agricultural Biotechnology: Opportunities for International Development.* CAB International, Wallingford, UK.

Byrne, N.J. (1989) *The Scope of Intellectual Property Protection for Plants and Other Life Forms: A Report prepared for the Common Law Institute of Intellectual Property.* Intellectual Property Publishing Limited, London, UK.

Chilliard, J.S. (1988) Long-term effects of recombinant bovine somatotropin (rBST) on dairy cow performances: a review. Proceedings EEC Seminar, Brussels, Belgium, September 1988.

Chilliard, J.S. and Soller, M. (1987) Molecular markers in the genetic improvement of farm animals. *Biotechnology*, 5, 573–6.

Cohen, J. (1989) Biotechnology research for the developing world. *Trends in Biotechnology*, 7 (11), 295–303.

Cunningham, E.P. (1990) Animal production. In: Persley, G.J. (ed.), *Agricultural Biotechnology: Opportunities for International Development*. CAB International Wallingford, UK.

Dale, J.L. (1990) Banana and plantain. In: Persley, G.J. (ed.), *Agricultural Biotechnology: Opportunities for International Development*. CAB International, Wallingford, UK.

Dart, P.J. (1990a) Plant production: introduction. In: Persley, G.J. (ed.), *Agricultural Biotechnology: Opportunities for International Development*. CAB International, Wallingford, UK.

Dart, P.J. (1990b) Agricultural microbiology: introduction. In: Persley, G.J. (ed.), *Agricultural Biotechnology: Opportunities for International Development*. CAB International, Wallingford, UK.

Dodds, J.H. and Tejeda, M. (1990) Potato. In: Persley, G.J. (ed.), *Agricultural Biotechnology: Opportunities for International Development*. CAB International, Wallingford, UK.

Doyle, J.J. and Spradbrow, P.B. (1990) Animal health. In: Persley, G.J. (ed.), *Agricultural Biotechnology: Opportunities for International Development*. CAB International, Wallingford, UK.

Duvick, D.N. (1990) Plant breeding. In: Persley, G.J. (ed.), *Agricultural Biotechnology: Opportunities for International Development*. CAB International, Wallingford, UK.

EC (1988) *Proposal for a Council Directive on the Legal Protection of Biotechnological Inventions*. Commission of the European Community, Brussels, Belgium, 17 October 1988.

EPO (1988) *Patenting Life Forms*. European Patent Office, The Hague, The Netherlands.

Evenson, R.E. and Putman, J. (1990) Intellectual property management. In: Persley, G.J. (ed.) *Agricultural Biotechnology: Opportunities for International Development*. CAB International, Wallingford, UK.

Farrington, J. (ed.) (1989) *Agricultural Biotechnology Prospects for the Third World*. Overseas Development Institute, Regents College, Regents Park, London.

Fielder, D.R., Spradbrow, P.B. and Dart, P.J. (1990) Aquaculture. In: Persley, G.J. (ed.), *Agricultural Biotechnology: Opportunities for International Development*. CAB International, Wallingford, UK.

Fishlock, D. (1989) Biotechnology survey. *Financial Times* (London), 12 May 1989, 33–5.

Fowler, C., Lachkovics, E., Mooney, P. and Shand, H. (1988) The Laws of Life: Another Development and the New Biotechnologies. *Development Dialogue*, 1–2, 1–350.

Giddings, L.V. (1990) Microbial bioprocessing. In: Persley, G.J. (ed.), *Agricultural Biotechnology: Opportunities for International Development*. CAB International, Wallingford, UK.

Goodman, D., Sorj, A. and Wilkinson, J. (1983) *From Farming to Biotechnology*. Blackwell, Oxford, UK.

Hanrahan, T.J. (1988) Use of somatotropin in livestock production: growth in pigs. Proceedings EEC Seminar, Brussels, Belgium, September 1988.

Herdt, R.W. and Riley, F.Z. (1987) International rice research priorities: implications for biotechnology initiatives. In: *Proceedings Rockefeller Foundation*

Workshop on Allocating Resources for Developing Country Agricultural Research, Bellagio, Italy, 6–10 July 1987. Rockefeller Foundation, New York.

Holloway, B. (1990) Educational needs. In: Persley, G.J. (ed.), *Agricultural Biotechnology: Opportunities for International Development.* CAB International, Wallingford, UK.

IBPGR (1988) *Annual Report.* International Board for Plant Genetic Resources, Rome, Italy.

IFPRI (1988) *Annual Report.* International Food Policy Research Institute, Washington, DC.

Jaenisch, R. (1988) Transgenic animals. *Science,* 240, 1468–73.

James, C. and Persley G.J. (1990) Role of the private sector. In: Persley, G.J. (ed.), *Agricultural Biotechnology: Opportunities for International Development.* CAB International, Wallingford, UK.

Javier, E. (1990) Issues for national agricultural research systems. In: Persley, G.J. (ed.), *Agricultural Biotechnology: Opportunities for International Development.* CAB International, Wallingford, UK.

Jones, D.A., Ryder, M.N., Clare, R.G., Farrand, S.K. and Kerr, A. (1988) Construction of a Tra-deletion mutant of PAg K84 to safeguard the biological control of crown gall. *Molecular and General Genetics,* 212, 207–14.

Jones, K.A. (1990) Classifying biotechnologies. In: Persley, G.J. (ed.), *Agricultural Biotechnology: Opportunities for International Development.* CAB International, Wallingford, UK.

Jones, L.H. (1990) Perennial vegetable oil crops. In: Persley, G.J. (ed.), *Agricultural Biotechnology: Opportunities for International Development.* CAB International, Wallingford, UK.

Kenny, M. and Buttel, F. (1985) Biotechnology: prospects and dilemmas for Third World development. *Development and Change,* 16, 61.

Kerr, A. (1989) Commercial release of a genetically engineered bacterium for the control of crown gall. *Agricultural Science,* November, 41–4.

Krugman, S.L. (1990) Forestry. In: Persley, G.J. (ed.), *Agricultural Biotechnology: Opportunities for International Development.* CAB International, Wallingford, U.K.

Larkin, P.J. (1990) Wheat. In: Persley, G.J. (ed.), *Agricultural Biotechnology: Opportunities for International Development.* CAB International, Wallingford, UK.

Martin, B., Nienhuis, J., King, G. and Schaeffer, A. (1989) Restriction fragment length polymorphisms associated with water use efficiency in tomato. *Science,* 243, 1725–8.

Meeusen, R.L. (1990) Insect and disease resistance. In: Persley, G.J. (ed.), *Agricultural Biotechnology: Opportunities for International Development.* CAB International, Wallingford, UK.

Miller, S.A. and Williams, R.J. (1990) Agricultural diagnostics. In: Persley, G.J. (ed.), *Agricultural Biotechnology: Opportunities for International Development.* CAB International, Wallingford, UK.

Millis, N.F. (1990) Regulating release of organisms. In: Persley, G.J. (ed.), *Agricultural Biotechnology: Opportunities for International Development.* CAB International, Wallingford, UK.

NAS (1987) *Introduction of Recombinant DNA-Engineered Organisms into the Environment: Key Issues.* National Academy Press, Washington, DC.

Novak, F.J., Afza, R., Van Duren, M., Perea, M., Cogner, B.V. and Xiaolang, T. (1989). Somatic embryogenesis and plant regeneration in suspension cultures of dessert (AA and AAA) and cooking (AAB) bananas (*Musa* spp.). *Bio/Technology,* 7(2), 154–9.

NRC (1989) *Field Testing Genetically Modified Organisms. Framework for Decisions.* National Academy Press, Washington, DC.

OECD (1986) *Recombinant DNA Safety Considerations.* OECD, Paris.

OECD (1988) *Biotechnology and the Changing Role of Government.* OECD, Paris.

OECD (1989) *Biotechnology Economic and Wider Impacts.* OECD, Paris.

OTA (1984) *Commercial Biotechnology: An International Analysis, OTA-BA-218.* US Government Printing Office, Washington, DC.

OTA (1986) *Technology Public Policy and the Changing Structure of American Agriculture, OTA-F-285.* US Government Printing Office, Washington, DC.

OTA (1988a) *New Developments in Biotechnology: US Investments in Biotechnology, OTA-BA-360.* US Government Printing Office, Washington, DC.

OTA (1988b) *Field Testing of Engineered Organisms: Genetic and Ecological Issues. OTA-BA-350.* US Government Printing Office, Washington, DC.

OTA (1989) *New Developments in Biotechnology: Patenting Life. Special Report, OTA-BA-370.* US Government Printing Office, Washington DC.

Persley, G.J. (1989) The application of biotechnology to agriculture in developing countries. *Agbiotech News and Information,* 1, 23-6.

Persley, G.J. (1990a) *The Coconut Palm: Prosperity or Poverty?* Consultative Group on International Agricultural Research Technical Advisory Committee, FAO, Rome. 81pp.

Persley, G.J. (1990b) *Coconut Research Opportunities.* Consultative Group on International Agricultural Research Technical Advisory Committee, FAO, Rome.

Persley, G.J. (1990c) Policy options for international development agencies. In: Persley, G.J. (ed.), *Agricultural Biotechnology: Opportunities for International Development.* CAB International, Wallingford, UK.

Persley, G.J. and Peacock, W.J. (1990) Biotechnology for bankers. In: Persley, G.J. (ed.), *Agricultural Biotechnology: Opportunities for International Development.* CAB International, Wallingford, UK.

Plucknett, D., Cohen, J.I. and Horne, M. (1990) Role of the international agricultural research centres. In: Persley, G.J. (ed.), *Agricultural Biotechnology: Opportunities for International Development.* CAB International, Wallingford, UK.

Pritchard, A.J. (1990) Lending by the World Bank for agricultural research: a review of the years 1981 through 1987. *World Bank Technical Paper Number 118,* World Bank, Washington, DC.

Quirke, J.F. and Schmid, H. (1988) Application of biotechnology to animal production. Proceedings VI World Conference on Animal Production, Helsinki, Finland, June 1988.

Raff, J. (1990) Infrastructural requirements. In: Persley, G.J. (ed.), *Agricultural Biotechnology: Opportunities for International Development.* CAB International, Wallingford, UK.

Randles, J. and Hanold, D. (1988) *Annual Report, Coconut Virus Diseases Project.* Australian Centre for International Agricultural Research. Canberra, Australia.

RDMC (1987) *Recombinant DNA Monitoring Committee Publication No. 7. Procedures for the Assessment of the Planned Release of Recombinant DNA Organisms. May 1987.* Department of Industry, Technology and Commerce, Canberra, Australia.

Scowcroft, W.R. (1990) Annual vegetable oil crops. In: Persley, G.J. (ed.), *Agricultural Biotechnology: Opportunities for International Development.* CAB International, Wallingford, UK.

Söndahl, M.R. (1990) Coffee and cocoa. In: Persley, G.J. (ed.), *Agricultural*

Biotechnology: Opportunities for International Development. CAB International, Wallingford, UK.

Sussman, M., Collins, G.H., Skinner, F.A. and Stewart-Tull, D.E. (eds) (1988) *The Release of Genetically-Engineered Micro-Organisms.* Academic Press, London.

Tanksley, S.D., Young, N.D., Paterson, A.H. and Bonierbale, N.W. (1989) RFLP mapping in plant breeding: new tools for an old science. *Biotechnology,* 7(3), 257–64.

Tiedje, James, M., Colwell, R.K., Grossman, Y.L., Hodson, R.E., Linski, R.E., Mack, R.N. and Regal, P.J. (1989) The planned introduction of genetically engineered organisms: ecological considerations and recommendations. *Ecology,* 70, 298–315.

Toennison, G.H. (1990) Rice biotechnology: progress and prospects. In: *Proceedings of Conference on Pest Management in Rice, 6–9 June, 1990.* Rockefeller Foundation, New York.

Toennison, G.H. and Herdt, R.W. (1989) The Rockefeller Foundation's international program on rice biotechnology. In: *Proceedings USAID Workshop on Strengthening Collaboration in Biotechnology: International Agricultural Research and the Private Sector, 17–21 April 1988.* Science and Technology Bureau, USAID, Washington, DC.

USAID (1989) *Proceedings of a Workshop on Strengthening Collaboration in Biotechnology: International Agricultural Research and the Private Sector, 17–21 April 1988.* Science and Technology Bureau, USAID, Washington, DC.

USDA (1988) *Plant Genome Research Conference Report.* United States Department of Agriculture, Washington, DC.

Van Brunt, J. (1987) Bringing biotech to animal health care. *Biotechnology,* 5, 677–83.

Ward, K.A. (1988) Transgenic animals. *Proceedings Course on Molecular Genetics Applied to Animal Breeding and Production.* National Research Council, Ottawa, Canada.

Whitten, M.J. and Oakeshott, J.G. (1990) Biocontrol of insects and weeds. In: Persley, G.J. (ed.), *Agricultural Biotechnology: Opportunities for International Development.* CAB International, Wallingford, UK.

Wilmut, I., Clark, J. and Simons, P. (1988) A revolution in animal breeding. *New Scientist,* 7 July, 56–9.

World Bank (1989) *Agricultural Biotechnology Study.* World Bank. Technical Report, December, The World Bank, Washington, DC.

Wright, B. (1989) Gene-spliced pesticide uncorked in Australia. *New Scientist,* 4 March, 23.

Wyke, A. (1988) The genetic alternative: a summary of biotechnology. *The Economist,* 30 April.

Yilma, T., Hsuy, D., Jones, L., Owens, S., Grubman, M., Mebus, D., Yamanaka, M. and Dale, B. (1988) Protection of cattle against rinderpest with vaccinia virus recombinants expressing the HA or F gene. *Science,* 242, 1058–61.

Glossary of terms

Amino acid: Any one of a group of 20 chemicals that are linked together in various combinations to form *proteins*. Each protein is made up of a specific sequence of these chemicals. This unique sequence is coded for by a *gene*.

Anticodon: A particular combination of three *bases* in *transfer RNA* that is *complementary* to a specific three-base *codon* in *messenger RNA*. Alignment of codons and anticodons is the basis for organising amino acids into a specific sequence in a *protein* chain.

Bacterium: Any of a group of one-celled microorganisms having round, rodlike, spiral or filamentous bodies that are enclosed by a cell wall or membrane and lack fully differentiated nuclei.

Base: The units of nucleic acids. In DNA, the four bases are adenine (A), guanine (G), cytosine (C) and thymine (T). In RNA, the base uracil (U) replaces thymine. Bases are sometimes called nucleotides.

Base pairing rule: Two bases, one in each strand of a double-stranded DNA molecule, are attracted to one another on the basis of their chemical structure, so that G (guanine) always pairs with C (cytosine), and A (adenine) pairs with T (thymine) in DNA or U (uracil), in RNA. Thus by knowing the sequence of bases in one strand of DNA, it is possible to predict the sequence in the opposite, complementary strand.

Biotechnology: Any technique that uses living organisms or substances from those organisms to make or modify a product, to improve plants or animals, or to develop microorganisms for specific uses. These techniques include the use of new technologies such as recombinant DNA, cell fusion and other new bioprocesses.

Cell: The smallest component of life. A membrane-bound protoplasmic body capable of carrying on all essential life processes. A single cell unit is a complex collection of molecules with many different activities.

Chromosome(s): The physical structure(s) within a cell's nucleus, composed of a DNA-protein complex, and containing the hereditary material i.e. genes; in bacteria the DNA molecule is a single closed circle (without protein).

DNA (deoxyibonucleic acid): The molecule that is the repository of genetic information in all organisms (with the exception of a few viruses). The information coded by DNA determines the structure and function of an organism.

DNA sequencing: Determination of the order of bases in a DNA molecule.

Enzyme: A *protein* which accelerates a specific chemical reaction, without itself being destroyed.

Gene: The fundamental physical and functional unit of heredity, the portion of a DNA molecule that is made up of an ordered sequence of nucleotide base pairs that produces a specific product or have an assigned function.

Genetic code: The code that translates information contained in *messenger RNA* into *amino acids*. Different triplets of bases (called *codons*) code for each of 20 different amino acids.

Genetic engineering: Technologies (including recombinant DNA technologies) used by scientists to isolate genes from an organism, manipulate them in the laboratory and insert them into another organism.

Genotype: The genetic constitution of an organism as distinguished from its physical appearance (phenotype).

Germplasm: The total genetic variability, represented by germ cells or seeds, available to a particular population of organisms.

Hybrid: An offspring of a cross between two genetically unlike individual plants or animals.

Hybridoma: A new cell resulting from the fusion of a particular type of immortal tumor cell line, a myeloma, with an antibody-producing B lymphocyte. Cultures of such cells are capable of continuous growth and specific (i.e. monoclonal) antibody production.

Intellectual property: That area of the law involving patents, copyrights, trademarks, trade secrets and plant variety protection.

Ligase: An enzyme that joins the ends of DNA molecules together. These *enzymes* are essential tools in genetic engineering.

Monoclonal antibodies (MCA): Identical antibodies that recognise a single, specific antigen and are produced by a clone of specialised cells.

Recombinant DNA: Hybrid DNA sequences assembled *in vitro* from different sources; or hybrid DNA sequences from the same source assembled *in vitro* in a novel configuration.

Restriction enzymes: Certain bacterial enzymes that recognise specific short sequences of DNA and cut the DNA where these sites occur.

RFLP: Restriction fragment length polymorphism. Fragments of differing lengths of DNA that distinguish individuals, produced by cutting with restriction enzymes. They result from DNA sequence variations and can be detected with radioactive probes and used as markers in breeding.

RNA (ribonucleic acid): Nucleic acid complementary to DNA. The three kinds of RNA important in the genetic processes in cells are messenger RNA (mRNA), ribosomal RNA (rRNA) and transfer RNA (tRNA).

Species: Reproductive communities and populations that are distinguished by their collective manifestation of ranges of variation with respect to many different characteristics and qualities.

Tissue culture: The propagation of tissue removed from organisms in a laboratory environment that has strict sterility, temperature and nutrient requirements.

Transcription: The process of converting information in *DNA* into information contained in *messenger RNA*.

Transfer RNA (tRNA): *RNA* that is used to position *amino acids* in the correct order during protein construction.

Transformation: Introduction and assimilation of DNA from one organism to another via uptake of naked DNA.

Transgenic animals or plants: Animals or plants whose hereditary DNA has been augmented by the addition of DNA from a source other than parental germplasm, in a laboratory using recombinant DNA techniques.

Translation: The process of converting the information in *messenger RNA* into protein.

Vector: A carrier or transmission agent. In the context of recombinant DNA technology, a vector is the DNA molecule used to introduce foreign DNA into host cells. Recombinant DNA vectors include plasmids, bacteriophages and other forms of DNA.

Sources: OTA (1988, 1989).

Acronyms and abbreviations

ACIAR	Australian Centre for International Agricultural Research
AIDAB	Australian International Development Assistance Bureau
CEC	Commission of the European Communities
CGIAR	Consultative Group on International Agricultural Research
CIAT	Centro Internacional de Agricultura Tropical
CIMMYT	Centro Internacional de Mejoramiento de Maiz y Trigo
CIP	Centro Internacional de la Papa
EPO	European Patent Office
FAO	Food and Agriculture Organization
GATT	General Agreement on Tariffs and Trade
IARC	International Agricultural Research Centre
IBC	Institutional Biosafety Committee
IBPGR	International Board for Plant Genetic Resources
ICARDA	International Center for Agricultural Research in the Dry Areas
ICRISAT	International Crops Research Institute for the Semi-Arid Tropics
ICSU	International Council of Scientific Unions
IFC	International Finance Corporation
IITA	International Institute of Tropical Agriculture
ILCA	International Livestock Centre for Africa
ILO	International Labour Office
ILRAD	International Laboratory for Research on Animal Diseases
INIBAP	International Network for the Improvement of Banana and Plantain
IPR	Intellectual Property Rights
IRHO	Institut de Recherche pour les Huiles et Oléagineux
IRRI	International Rice Research Institute
ISNAR	International Service for National Agricultural Research
ISO	International Standards Organisation

NARS	National Agricultural Research System
NAS	National Academy of Science (USA)
NRC	National Research Council (USA)
OECD	Organization for Economic Co-operation and Development
ORSTOM	Office de Recherche Scientifique et Technique Outremer (France)
OTA	Office of Technology Assessment of the US Congress
PCT	Patent Co-operation Treaty
UNEP	United Nations Environment Programme
UNIDO	United Nations Industrial Development Organization
UPOV	International Union for the Protection of New Varieties of Plants
USAID	United States Agency for International Development
USDA	United States Department of Agriculture
WHO	World Health Organization
WIPO	World Intellectual Property Organization

Index